BACTERIA AND BAYONETS

BACTERIA
AND
BAYONETS
THE IMPACT OF DISEASE IN
AMERICAN MILITARY HISTORY

DAVID R. PETRIELLO

CASEMATE
Philadelphia & Oxford

Published in the United States of America and Great Britain in 2016 by
CASEMATE PUBLISHERS
1950 Lawrence Road, Havertown, PA 19083
and
10 Hythe Bridge Street, Oxford, OX1 2EW

ISBN 978-1-61200-341-2
Digital Edition: ISBN 978-1-61200-342-9

Cataloging-in-publication data is available from the Library of Congress and
the British Library.

10 9 8 7 6 5 4 3 2 1

Printed and bound in the United States of America.

For a complete list of Casemate titles please contact:

CASEMATE PUBLISHERS (US)
Telephone (610) 853-9131, Fax (610) 853-9146
E-mail: casemate@casematepublishing.com

CASEMATE PUBLISHERS (UK)
Telephone (01865) 241249, Fax (01865) 794449
E-mail: casemate-uk@casematepublishing.co.uk

CONTENTS

PREFACE   7

INTRODUCTION   9

One   Columbus Day or Contagion Day:
       Disease "Arrives" in America   11

Two   "Deus Flavit Nasus et Dissipati Sunt":
       The Protestant Wind and the Catholic Flu   23

Three Pocahontas and the Plague: *The English
       and Disease in the Conquest of the Colonies*   29

Four   "The Paths to Glory Lead but to
       the Grave": *Disease in the Early French
       and Indian Wars*   42

Five   "Pestilence Gave Them a Common
       Death": *Disease and the English Conquest
       of North America*   64

Six   Typhus and Taxation: *Disease and the
       American Revolution*   74

Seven A Nation Forged in Gout and
       Expanded by Venereal Disease:
       A Medical Look at the Early Republic   116

Eight Montezuma's Revenge: *Disease and
       Manifest Destiny*   138

Nine   Johnny Dysentery and Billy Typhus:
       Disease and the Civil War   151

Ten   Remember the *Maine*, to Hell with
       Yellow Fever: *Imperialism and Illness*   172

Eleven   Love in the Age of Cholera,
Warfare in the Age of Typhoid:
Progressivism and Pestilence   185

Twelve   Bullets, Bayonets, and Botulism:
Biological Warfare in the Twentieth Century   199

Thirteen   Al-Qaeda, Anthrax, and America:
*Terrorism and Disease in Post-Cold
War America*   219

CONCLUSION   225

ENDNOTES   227

BIBLIOGRAPHY   244

INDEX   259

PREFACE

What killed the most soldiers during the Civil War, what allowed the Spanish to conquer various Native American cultures, what doomed the British in the South during the American Revolution, what prevented America from conquering Canada . . . disease. Having studied and taught American history for years, it became evident to both my students and myself that the answer to many of the questions regarding American historical events was *disease*. Surely if the term continued to occur as one progressed through a study of the country's past, it must be an important factor throughout the nation's military history. Thus, this work was created to more carefully study and detail that specific importance and detail how history has often hinged upon microbes.

In many ways though, this book is a study of "what-ifs." When one studies history, the impact of an event is the prime focus, yet the possibilities of what could have been should also be kept in mind. It is historical fact that disease destroyed many Native American civilizations, pestilence wiped out the Viking colonies of Greenland, and that the American invasion of Canada in 1775 was stopped by illness. What if disease had not wiped out the Native Americans, what if the Viking colonies had not succumbed to plague, what if a variety of illnesses had not crippled the American attack on Quebec? This book seeks to not only examine the interconnectedness between disease and American military history, but also challenges the reader to think of how the nation and world would have been different had the various pestilences not arisen and impacted events when and how they did. As history and experience have shown time and time again, much of war cannot be planned for.

Yet, this book does not seek to establish that disease was the *only* factor that contributed to significant events in American military history. Historical events are the products of a series of causes and variables. While some historians devote both books and their lifetimes to promoting one or another factor as the *sine qua non*, this author merely seeks to address one of the many factors in the evolution of various American historical events, not elevate it to primary status. That decision is left up to the reader.

In the matter of an apologia, the history of disease is an immense topic. Books could be written alone on each chapter, or subchapter of this work. The author does not claim to provide here a history of all disease in the chronology of America. This book should merely be seen as an overview or snapshot of the impact that disease has had on the nation's development. An introduction to a field of study underappreciated and in need of further examination.

INTRODUCTION

"War is the father of all."
—HERACLITUS

W hen Heraclitus opined the above statement, he was perhaps making a broader claim about the evolution of both man and society. On any scale, competition drives advancement. For Heraclitus, war was the ultimate form of competition; the greatest change agent in the ancient world. War produced heroes, leaders, empires, tools, inventions, exploration—in short, advancement. Yet for the ancient world in general, the Greeks in particular, and even Heraclitus personally, there was a greater factor impacting life and society: disease.[1]

The ancients undoubtedly appreciated disease as a powerful force in life. Such major deities as Apollo and Isis were duly consulted during outbreaks of contagion by the Greeks and the Egyptians. In addition, plague was a major plot tool utilized by the poet Homer at the beginning of the martial classic the Iliad, where a vengeful Apollo Smintheus cut down the Greeks with his arrows of pestilence. Finally, to appreciate the impact of disease upon ancient military history, one needs look no further than the Periclean Plague that swept Athens early in the Peloponnesian War, or the Justinian Plague that doomed the impending re-conquest of Italy by the Byzantine Empire. Though the nature of these contagions is still debated, their effects are well established. The first pestilence not only carried off Athens' most promising leader but severely crippled its ability to wage war as well, while the latter accomplished what hordes of barbarians could not. In fact, it would not be hyperbole to credit these two sicknesses as major factors in both Athens' subsequent loss to Sparta and the permanent dissolution of Europe.

History is replete with examples of how contagion impacted war and battles on every continent. Evidence of this arises in the Black Plague that was pushed westward by the Mongol invasions, the downfall of Alexander's great empire after he succumbed to contagion, and the typhus that crippled the armies of Napoleon in Russia, not "General Winter" which is normally blamed. How many thousands of other battles, skirmishes, campaigns, and even wars were in some part impacted by the presence of illness? In fact siege as a very tactic had at its core the desire to strangle an enemy until famine and disease reduced him to surrender.

Appreciating its influence, many famous authors of military canons included the issue of pestilence within their works. Sun Tzu, for one, advised commanders to, "*camp on hard ground, the army will be free from disease of every kind, and this will spell victory.*"[2] This is an obvious reference to the perceived presence of disease in swamps or marshy areas under the old miasma theory. Across the Eurasian continent, Flavius Vegetius devotes an entire section of his book to the health of soldiers. He likewise argues against camping in low lying or marshy areas, but then goes further to caution against marching in cold weather and consuming bad water. "*It is hard for those who are fighting both a war and disease.*"[3] The Roman commander urges his readers to remedy the situation by engaging the soldiers in daily exercises to keep them fit and less susceptible to illness, very modern thinking indeed. Men from Machiavelli to Napoleon to Adolf Hitler likewise opined on the effects of disease on the military might of a nation.

Contagion has historically impacted every segment of warfare. It lurks in the causes, drives planning, preparation, and strategy, and ultimately helps to decide the winner. It is as true in American military history as that of any other nation. If Heraclitus is correct, and war is the father of all, than disease is the mother.

CHAPTER ONE

COLUMBUS DAY OR CONTAGION DAY
DISEASE "ARRIVES" IN AMERICA

"A pestilence seized them, characterized by great pustules, which rotted their bodies with a great stench, so that their limbs fell to pieces in four or five days."
—DIEGO DE LANDA, 16TH CENTURY

I n May of 1521 Hernan Cortes stood before the sprawling and well protected Aztec capital of Tenochtitlan, one of the largest cities in the world at the time. The Spanish had been violently expelled from the city the year before during the infamous *La Noche Triste* (Night of Sorrows). Yet Cortes now brazenly besieged the metropolis of perhaps 100,000 souls with an army of only slightly over a thousand Spaniards.[1] What followed was one of the most remarkable military accomplishments of the last two thousand years. In less than three months, over half of the Aztec population was dead, along with their emperor, and the once great capital was reduced to smoldering ruins. The Aztec Empire, a regional juggernaut a generation before, was laid low.

The military history of America began with the military conquest of the New World by Spain and the other great European powers. Their tactics used against the Natives and the importance of pestilence to their ultimate victory would set the standard for future actions by the United States. Beyond simply the conquest of the Aztecs, the arrival of the Spanish in the Americas brought about a series of victories unprecedented in military history. A collection of small conquistador-led armies managed to destroy a number of major indigenous empires from the Mississippi to Patagonia that rivaled some European kingdoms in size and population. As well, hundreds of smaller societies and tribes from California to Tierra del Fuego

11

became subject to the Spanish Empire. This conquest has been attributed to a variety of advantages on the part of the Europeans, including their possession of more advanced weapons and materials, their use of horses, and their effective utilization of discontented native allies.[2] Yet the single most important factor was disease, a weapon unplanned for by the Spanish and unappreciated by history until just the last century. Former theories of Spanish brutality, better known as the Black Legend and popularized by the works of De Las Casas among others, as the main cause of Native depopulation have been largely discredited in favor of a mass extinction resulting from the various diseases introduced unwittingly by the conquistadors.[3] In fact, estimates of depopulation among the native peoples of the Americas from disease range as high as 90% of their pre-Columbian population. Without pestilence the Spanish would not have been able to depopulate, conquer, and settle in the patterns and timescale that they did. The conquest of the New World by the Old was accomplished more on the backs of viruses and bacteria than on horses.

Examining Cortes' conquest of Tenochtitlan more deeply it becomes evident that his victory was largely made possible due to the smallpox outbreak that erupted following the expulsion of the Spanish during *La Noche Triste*. From the entrance of Cortes and his men into the Aztec island capital in November of 1519 until their expulsion by force in July of the next year, the Spaniards had ample opportunities to spread disease amongst the population of the metropolis.[4] The contagion lasted until the winter of 1520 and claimed well over 25% of the city's population.[5] As the death rate climbed houses were simply demolished to bury the bodies inside, while the corpses of thousands of others floated among the *chinampas* (island gardens) of the city. More devastating for the Aztec resistance was the death of Montezuma's successor, Cuitlahuac, combined with a large number of the Aztec elite; the very leaders who would have been responsible for resisting the Spanish conquest.[6] In effect the population of the capital was already demoralized and devastated when Cortes returned in the late spring of 1521 with his makeshift army.[7] In addition, the subsequent siege disrupted both food and water supplies to the capital, leading to further famine-related illnesses. *"The ground and lake and fighting platforms were all full of dead bodies; they smelled so bad that there was no man who could bear it . . . and even Cortes was sick from the stench that penetrated into his nostrils."*[8] By August of 1521 the city had fallen, and Cortes' allies rampaged through the deserted and dev-astated streets enslaving or evicting those who remained. The great city of Tenochtitlan had fallen as much through pestilence as through force of arms. This should not be construed though to belittle the martial skill and planning of Cortes, which should be appreciated, but only to place it more into context.

Similar stories unfolded throughout the Americas during the first half of the 16th century. When Columbus first landed in the Caribbean, the local Taino population is estimated to have been around 500,000 souls. Over the course of the next thirty years 85–90% of their pre-Columbian population was lost. The overwhelming majority of these deaths resulted from disease, not as has been historically argued, the brutality of the Span-ish. *"There occurred an epidemic of smallpox so virulent that it left Hispaniola, Puerto Rico, Jamaica, and Cuba desolated of Indians."*[9] During the second decade of the 16th century, Spanish attempts to settle Panama were similarly visited by the specter of pestilence. In 1514 over 700 settlers at Darien succumbed to malaria and yellow fever, eventually leading to the overthrow of the duly appointed crown governor in favor of the explorer Vasco Núñez de Balboa. Gonzalo Fernández de Oviedo y Valdés estimated in his work *Writing from the Edge*, based upon his own first hand observations, that over two million indigenous people in the Panama region succumbed to illness as well over the course of the next twenty years. Though the numbers may be exagger-ated, the overall picture is not. Conditions worsened to the point that by

Christopher Columbus receiving gifts from the Cacique Guacanagari, by Theodor de Bry.

1527 requests were dispatched to the Spanish crown proclaiming the need for slave laborers to replace the depleted native and Spanish populations. In fact the entire system of African slavery that would eventually come to dominate labor in the New World had as its main catalyst disease.

Other examples on the same scale as the Aztecs' experience with pestilence occurred within South America as well. Francisco Pizarro landed in Ecuador in 1532, ostensibly to explore the interior of the continent, make contact with the various native groups, and establish Spanish settlements. Having heard that the powerful Incan ruler known as Atahualpa was residing at the Incan Baths near Cajamarca, Pizarro began the two-month march towards him. As the Spanish neared they dispatched the friar Vincente de Valverde to negotiate with the natives and convert Atahualpa. According to popular accounts, after de Valverde had approached the Inca with a copy of his breviary, the proud god king threw it to the ground. Outnumbering the Spanish by over three hundred to one, Atahualpa undoubtedly thought he had little to fear as he was carried upon his litter into the town of Cajamarca, believing it to be folly to be concerned with so few men. Yet within the course of a few minutes, Francisco Pizarro with only 168 men had managed to kill over 2,000 Inca and capture the divine emperor himself. What followed was the quick dissolution and conquest of one of the world's largest empires at the time.[10] Yet to understand the success of the Spanish we need to travel back in time five years.

In 1527 the grandson of the Incan Empire's founder sat securely on his throne. Huayna Capac had done his own part to further expand the empire, stretching it as far south as Chile and Argentina. Yet he began to hear reports of a foreign group which had landed in the far north of his empire and duly

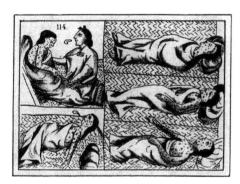

16th century drawing of Aztec smallpox victims from the Florentine Codex.

proceeded towards the area of modern day Colombia. Though he campaigned actively in the region, he never encountered the Spanish or any other foreign group, but in the end he did bring more than plunder back to Cuzco with him. Huayna Capac and many other Inca had become infected. There remains debate as to whether the disease which befell the Inca was smallpox, or bartonellosis, or some other illness.[11] Regardless, by the end of 1527 the emperor, his heir Ninan Cuyochi, and over a hundred thousand others had died. The disaster of this episode was that without a clear line of succession to the throne, civil war soon erupted between the remaining sons of Huayna Capac. A vicious internecine war raged from 1529 to 1532 between Huascar and Atahualpa. By the end of the struggle, an estimated 5% or more of the empire's population had been killed, a much higher percentage than experienced in the American Civil War. Fields were ruined, villages depopulated, loyalties were questioned, and an empire built upon conquest and enslavement was beginning to show cracks. It was at this precise waning of Incan power that the army of Pizarro arrived. Yet again an epidemic that had preceded the conquistadors became the catalyst for the destruction of another native empire.

The northernmost major empire at the time of discovery was the Mississippi Culture, a civilization that experienced a sudden and mysterious decline which still baffles historians to this day. By the time of Hernando de Soto's march through the American Southeast in 1539, the population centers of the Mississippians with their advanced mound structures and populations that rivaled then contemporaneous cities in Europe had been largely abandoned. De Soto encountered only a few Mound Builder centers, and these would themselves largely disappear over the next half century.[12] In fact, by the 18th century the various Native groups of the region had completely lost the skill and knowledge to build the mounds, producing various theories of the Mound Builders as a separate, often non-Native, population group.[13] A variety of theories have evolved as to why the Mississippi Culture declined; with such densely packed villages, could disease have played a factor?

Perhaps the greatest successes of the Spanish came from their rather small and unassuming exploratory expeditions that trekked across the southeastern and southwestern parts of the modern United States. Though small in size, these voyages altered the political and military landscape of the region by spreading numerous pathogens in their wake. Cabeza de Vaca's eight-year journey from Florida to Mexico City as part of the Narvaez Expedition, though itself decimated by disease, also helped to introduce many new contagions into the southern half of the United States, undoubtedly helping to pacify that region.

Nor were the Spanish the only beneficiary of disease in terms of conquering the New World. The Russian conquest of Alaska in the 18th century was in part made possible by the ravages of smallpox. The disease accompanied the Europeans as they methodically explored and settled the region. A particularly virulent outbreak in 1838 prompted the Eskimos to raid various Russian settlements the next year. Russia had to waste valuable resources building forts to fend off the Natives, slowly calling into question the century-old enterprise.

The reason for this microbial imbalance between the Old and New Worlds has been well established by such writers as Jared Diamond and others. Geographic determinism resulted in the inhabitants of Europe, Asia, and Africa possessing a variety of animals that were able to be domesticated over the course of 10,000 years. Zoonosis, or the transfer of disease from animals to man, contributed to the majority of deadly pestilences that were common in the Eastern hemisphere.[14] From smallpox to measles the domestication of animals brought disease into early society, while millennia of exposure produced a certain level of immunity, or at least familiarity.

Native Americans, with the exception of the llama and guinea pig, possessed no major animals to domesticate. Eurasians on the other hand possessed 13 of the 14 major domesticable animals on the planet. Relying on these beasts for food, labor, and transportation, Europeans, Asians, and Africans cohabited in close contact with their animals. The 10,000 years since the Neolithic Revolution thus provided ample time for animal diseases to mutate and transfer to humans. Though this did not result in a total level of immunity for man, it did at least lead to a practical level of familiarity and resistance. Thus, without the presence and/or domestication of these animals by the Paleo-Indians of the Americas, they were more susceptible to the various scourges brought over by the Europeans.

Yet, though *certain* epidemic diseases were unknown to Native Americans before the arrival of Columbus, this should not be taken to mean that *no* disease was present in the Western Hemisphere.[15] Some of the more common illnesses that have been discovered among pre-Columbian Indians include iron deficiency anemia, osteoarthritis, tuberculosis, herpes, Carrion's Disease or bartonellosis, and hepatitis.[16] Not as epidemic or acute as Old World contagions, these New World sicknesses tended to be chronic and geographically contained.[17] Yet it is safe to argue that Native Americans were not living in the carefree, germ-free paradise often associated with them. In fact more recent scholarship has begun to challenge even previously held assumptions about the non-presence of epidemic disease within the Western Hemisphere.

Recently a few historians and scientists have begun to question the possibility of whether or not larger epidemic disease was already present in some American cultures prior to the arrival of the Spanish as well. Though Native cultures may have avoided the zoonosis associated with Eurasian domestication, they may have been exposed to diseases arising from and vectored by other sources. This includes the lice borne pestilence of typhus.[18] According to Felipe Guaman Poma in his 17th-century *Chronicle*, it was thanks to a massive typhus outbreak that the Inca were able to conquer the region of Chile during the reign of Pachacuti Inca Yupanqui. *"The defeat of Chile was made possible by the ravages of plague . . . disease and famine, even more than force of arms brought about the downfall of the Chileans."*[19] This stands as a striking foreshadowing to the fate of the Inca themselves almost a century later. Yet the most controversial of these preexisting epidemic diseases remains the mysterious *Huey Cocoliztli*.[20] First described by both Fray Juan de Torquemada and the Spanish court physician Dr. Francisco Hernandez in the second half of the 16th century, this disease apparently ravaged central Mexico in both 1545 and 1576.

In Fray Torquemada's own words describing the 1576 outbreak:

In the year 1576 a great mortality and pestilence that lasted for more than a year overcame the Indians. It was so big that it ruined and destroyed almost the entire land. The place we know as New Spain was left almost empty. It was a thing of great bewilderment to see the people die. Many were dead and others almost dead, and nobody had the health or strength to help the diseases or bury the dead. In the cities and large towns, big ditches were dug, and from morning to sunset the priests did nothing else but carry the dead bodies and throw them into the ditches without any of the solemnity usually reserved for the dead, because the time did not allow otherwise. At night they covered the ditches with dirt . . . It lasted for one and a half years, and with great excess in the number of deaths. After the murderous epidemic, the viceroy Martin Enriquez wanted to know the number of missing people in New Spain. After searching in towns and neighborhoods it was found that the number of deaths was more than two million.

The outbreak resulted in an estimated 6 to 12 million deaths in 1545 and an additional 4 million in 1576. Taken together, these numbers would dwarf the estimated 8 million inhabitants of Mexico who died from the smallpox outbreak spread by Cortez's men in 1520. While for years it was assumed that *Cocoliztli* was simply another Western-introduced disease, the symptoms and overall mortality rates as pieced together from the writings of Dr. Hernandez, physician-in-chief of New Spain, cast new doubts upon this.

The fevers were contagious, burning, and continuous, all of them pestilential, in most part lethal. The tongue was dry and black. Enormous thirst. Urine of the colors sea-green, vegetal-green, and black, sometimes passing from the greenish color to the pale. Pulse was frequent, fast, small, and weak—sometimes even null. The eyes and the whole body were yellow. This stage was followed by delirium and seizures. Then, hard and painful nodules appeared behind one or both ears along with heartache, chest pain, abdominal pain, tremor, great anxiety, and dysentery. The blood that flowed when cutting a vein had a green color or was very pale [and] dry. . . . In some cases gangrene. . . invaded their lips, pudendal [genital] regions, and other regions of the body with putrefact members. Blood flowed from the ears and in many cases blood truly gushed from the nose. Of those with recurring disease, almost none was saved. Many were saved if the flux of blood through the nose was stopped in time; the rest died. Those attacked by dysentery were usually saved if they complied with the medication. The abscesses behind the ears were not lethal. If somehow their size was reduced either by spontaneous maturation or given exit by perforation with cauteries, the liquid part of the blood flowed or the pus was eliminated; and with it the cause of the disease was also eliminated, as was the case of those with abundant and pale urine. At autopsy, the liver was greatly enlarged. The heart was black, first draining a yellowish liquid and then black blood. The spleen and lungs were black and semi-putrefacted. . . the abdomen dry. The rest of the body, anywhere it was cut, was extremely pale. This epidemic attacked mainly young people and seldom the elder ones. Even if old people were affected they were able to overcome the disease and save their lives.[21]

From the above description the symptoms of the disease more resemble a tropical hemorrhagic fever than a Eurasian bacterial or viral infection. In addition, the very fact that it struck the younger and stronger members of societies rather than the elderly should identify this disease as something unique to the New World and more reminiscent of the Spanish Influenza outbreak in its fatality characteristics. The presence of one or more hemorrhagic fevers in the Americas predating the arrival of the Spanish is not as impossible as it may sound. Over the past half-century, over a dozen arenaviruses unique to the Americas have been discovered by scientists. Some of these being so virulent in fact, that they were chosen by the United States government to be used for its biological weapons program during the Cold War.[22] In addition, some of these arenaviruses employ rodents as vectoring agents, an animal present in large numbers in the New World. These mice tend to be self-domesticating, in that they historically have lived in close proximation with humans. Thus the advanced urban centers of Mesoamerica and South America could undoubtedly have been exposed to any number of these hemorrhagic fevers well before the arrival of the Spanish. Thus, though the indigenous people of the Americas did not domesticate large numbers of animals when compared to Eurasians, and therefore did not acquire resistance to disease in the classical sense, they were not completely disease free.

Historical studies have revealed that the population of Mexico and Central America fell from a peak of around 20 million in 1520 to around only one million a hundred years later in 1619. While the initial smallpox epidemic begun by the followers of Cortes in Tenochtitlan has been argued to have claimed 8 million lives, the additional 11+ million deaths over the course of the following century were largely assumed to have been the result of "other" diseases largely unidentified, or simply blamed on the "brutality" of the Spanish. Yet once the outbreaks of 1545 and 1576 are examined, a much deadlier epidemic than the smallpox wave of 1520 emerges. The mysterious outbreak recorded by the Spanish in 1545 bore an estimated 80% mortality rate, while that of 1576 produced a 45% death toll.

Historians have now generally accepted that the greatest tool in the conquistador's arsenal was disease; responsible for killing the majority of Amerindians and thus ensuring the Spanish conquest more than did guns, steel, or horses. The possibility of a hemorrhagic fever outbreak or additional local pestilence during the Conquest does not negate this theory but merely provides another interesting angle. Had the population of Mexico not been reduced by an additional 11 million lives following the Cocoliztli outbreaks would the handful of Spanish invaders and administrators present in the region been so successful? Would other diseases have possessed the power and potential to produce the same result? It seems from the evidence that Cocoliztli and the associated deer mouse, non-foreign invaders, appear to have brought about the downfall of the native people of Mexico and related peoples to an extent equal to or above the devastation caused by the Spanish and their germs. The Black Legend continues.

Finally, the very practices employed by the Native Americans in combating and treating disease proved to be worse than counterproductive when confronting the various pestilences of the 16th and 17th centuries. One of the most common treatments among the natives of North America involved sweat lodges. The individual would be placed in an enclosed building where a combination of bark and oils would be burned to "sweat" the sickness out of the individual in a way reminiscent of the European belief in humors. Unfortunately, as with modern day saunas, dehydration can develop if the steam process is not properly regulated. This would prove especially dangerous to individuals already suffering from such diuretic diseases as cholera or yellow fever. Following the time spent in the sweat lodge many cultures plunged the individual into cold water, a practice which could produce shock or heart failure in an already weakened patient.[23] Likewise, though Cortes reported advanced apothecaries within the Aztec capital, these peo-

Native Americans purging illness during Black Drink Ceremony, by Jacques le Moyne, 1564.

ples' overreliance on bleeding as a form of personal sacrifice undoubtedly presented a welcome invasion route for disease into the empire.

When individual efforts yielded no results, many Amerindians embraced tribal or religious solutions to the problems of disease. Besides standard dance and prayer ceremonies where the sick individual was brought into a general tribal powwow, many groups developed additional counter-productive religious rites. Numerous tribal bands employed fasting techniques to please the spirits that did little more than weaken tribal immune systems. The Cree Indians employed bear oil emetics during the 1780s to combat smallpox, which undoubtedly led to increased transmission. Henry Spelman and Cabeza de Vaca both reported native priests massaging diseased victims and attempting to suck out "poisons" from wounds.[24] In addition due to the close family band structure of most tribes, quarantining was not effectively practiced early on. On the contrary, village members would crowd into the homes of the infected to perform ceremonies or care for the sick. Roger Williams reported such a practice among the native peoples of Rhode Island, writing that when a native was sick his friends and family crowded around him, "*like a quire (sic) in prayer to their God for them.*" Even when quarantining became more widespread in the 18th century, it was done to the extreme, and many recoverable patients starved to death after being ostracized from the village. Once death did take place, burial practices by some cultures proved equally advantageous to the spread of disease. Thomas Harriot recorded in his *A Brief and True Report of the New Found Land of Vir-*

ginia how the deceased among the local Indians had their bodies opened, their flesh removed, and their bones dried. Undoubtedly during a time of epidemic this would have only catalyzed the spread of a disease.

Overall, the Native Americans became their own worst enemy in the fight against these invisible armies. Poor understanding of epidemiology, combined with counter-productive treatment methods, resulted in increased death rates at a time when European encroachment called for a healthy united front on the part of the Indians.

Regardless of whether disease preceded the Spanish to the Americas, or whether it arrived with them, the effects were the same. The various Amerindian empires of the New World collapsed due to relentless armies of microbes that they could not fight against. Had it not been for this invisible conquest, the reduction of the New World by Europeans would have taken far longer and involved far more resources and men. Disease laid low the New World and cleared the way for everything that was to follow. Just as the bubonic plague helped to usher in the societal and economic framework of the Renaissance, so too did the New World plagues set the groundwork for the colonies that followed.

CHAPTER TWO

"DEUS FLAVIT NASUS ET DISSIPATI SUNT"
THE PROTESTANT WIND AND THE CATHOLIC FLU

*"It would grieve any man's heart to see them
that have served so valiantly to die so miserably."*
—ADM. THOMAS HOWARD, 1588

The rise of English exploration and that nation's subsequent colonization of the New World is a story of bravery, perseverance, and luck. The Age of Discovery was already a century old when the English began to express interest in colonizing the Western Hemisphere. Yet standing between them and their future conquests was the mighty Habsburg Empire, a kingdom that spanned 13% of the planet's surface and included within its boundaries an equal amount of the world's people. During the last quarter of the 16th century, Spanish power was assumed in Europe, and Spanish domination was unchallenged in the New World. Yet due to a variety of vcircumstances, in a few short years Spanish dominion over Europe would wane, and its stranglehold upon New World colonization would collapse, paving the way for other nations to begin settling the Americas.

16th century Plague Doctor.

23

Through a combination of religious, political, and socio-economic factors, the kingdoms of England and Spain drifted towards collision in the last quarter of the 16th century. Stretching from 1585 until 1604 the war was waged in various theaters across the globe, from the Netherlands to Panama. Yet the most famous episode involved the launching of a direct attack by the Spanish Armada upon the British Isles. Spain dispatched the formidable Armada towards England in July of 1588 with the intention of landing an army from the Spanish Netherlands, defeating the English, converting the land, and incorporating the island into the Spanish Empire.

While at first glance the size and fame of the Spanish fleet was certainly intimidating, a closer examination reveals a variety of weaknesses in its composition and construction. The English ships were faster, more maneuverable, and contained guns of superior quality; the English crews were better trained in gunnery; and finally the Spanish ships were encumbered with over 180 priests, their assistants, and supplies. Yet victory for the English even with all of these advantages was far from a certainty.

English history has celebrated for hundreds of years the subsequent victory over the Armada. The story often begins with the fire ships of Calais, culminates with the English victory at Gravelines, and occasionally features the dénouement of the Armada foundering on the rocks off the coast of Ireland. Yet what is left out of the story is the role disease played in the Armada's destruction.

Our story actually begins on the voyage of the Armada from Lisbon to Cape Finisterre. Over the course of this 300-mile trip, disease and hunger began to wrack the fleet's crew. Duke Medina-Sidona himself in a letter to the King of Spain, Philip II, written in June reported the rapidly deteriorating conditions on the ships.

> Many of our largest ships are still missing . . . on the ships that are here there are many sick . . . these numbers will increase because of the bad provisions (food and drink). These are not only very bad, as I have constantly reported, but they are so scanty that they are unlikely to last two months. . . . Your Majesty, believe me when I assure you that we are very weak . . . how do you think we can attack so great a country as England with such a force as ours is now?[1]

Medina's poor provisions were made exponentially worse by the invcreased Spanish personnel on the ships as it prepared to sail. From February to June the number of men in the Armada swelled from 12,000 to 19,000. The Spanish commander thus decided to cut rations during the trip from Spain to the Netherlands, this decision would have made the crew even more vulnerable to disease during the dangerous voyage.

In the end it was a reduced Armada that passed through the English Channel. Yet disease at sea was an old nemesis, and was most likely not unexpected by the Spanish commanders. What was more disastrous though was the state of the Spanish *tercios* to be embarked by the Armada at Calais. As the fleet rounded the northern coast of France, Spanish forces in the Netherlands were being reduced from 30,000 battle-hardened men to roughly 16,000 by a variety of diseases. The invasion could be undertaken with a crew of weakened sailors, but it would go nowhere without an army. Even had the Armada survived the journey to Flanders, the condition of Parma's invasion force would have ultimately doomed the undertaking.

On July 27 the Spanish Armada anchored at Calais to await the embarkation of the disease depleted Spanish army. Before they could make that rendezvous, the English Sea Dogs attacked. While the English attack by fire ships sank none of the Spanish galleons, it did serve to break their formation, in which lay the Armada's strength. As the fleet drifted off the coast of Calais attempting to reform, the English navy finally struck. Five Spanish ships would be destroyed at the ensuing Battle of Gravelines. Though it was once again not an overwhelming tactical victory for England, Gravelines did produce certain instrumental strategic outcomes. First and foremost, the Spanish were postponed from completing their linkup with the *tercios* in the Spanish Netherlands. Worse yet, Spanish leadership became acutely aware that when they did manage the rendezvous, the condition of the invasion force would be less than that needed to conquer England. Combined, the defeat at Gravelines of the weakened Armada and the pestilence in the army camps spelled an end to the proposed invasion of the British Isles. In response the Duke of Medina-Sidonia decided to sail with his fleet back to Spain.

The concern for the Spanish Armada now became an issue of how best to make the voyage back from Calais to Santander. Rather than follow a reverse route back through the English Channel, where the Spanish risked

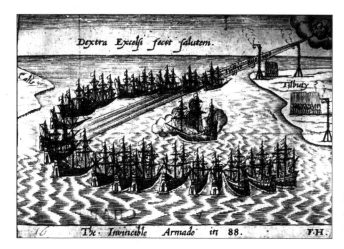

Spanish Armada and the Breath of God, 1588.

further attacks by the more accurate gunnery of the English navy, Medina-Sidonia chose the longer route around the British Isles. Sailing north and west around Scotland and Ireland, the Spanish hoped to avoid the English ships as well as receive aid and safe anchorage from the friendly, Catholic inhabitants of Ireland. Yet almost at once, the pre-existing conditions of disease and poor supplies began to take a toll on the Spanish fleet. More men succumbed to fatigue and famine, and many of the fleet's horses had to be thrown overboard due to a lack of fresh water and food.[2] Worse was yet to come.

Due to a combination of poor seamanship and poor luck, the Spanish Armada would be wracked by gales rounding the western coast of Scotland and Ireland. Not only did the wreckage of Spanish ships litter the beaches of the British Isles, but so did the corpses of 20,000 Spaniards as well. Two-thirds of the Spanish personnel died due to the storms as well as the various diseases that ravaged the ships and their crews along the way. When the *Trinidad Valancera* ran aground near Donegal on September 14, the crew decided to surrender to the English due to the extremes of hunger and disease they were experiencing, only to have 300 of their men slaughtered by their captors. Conditions in the Armada reached such a state of depredation that a Spanish force from the *Zuniga* landed near Liscannor Bay to obtain food after 80 of her men had died of hunger and disease. When the local Irish, erstwhile allies of the Spanish, refused to sell them goods, the crew resorted to force of arms to obtain the victuals.[3] Of the 10,000 men that returned to

Santander and various other harbors in Spain, a vast number would continue to perish over the next year due to the diseases and conditions of malnutrition that they brought back with them. Medina-Sidonia himself in fact blamed the diseases that had wracked his crew as the prime reason behind Spain's inability to defeat England.[4] On August 11th, the day after the Spanish began their long trek back home, Sidonia reported 3,000 cases of typhus aboard his ships, 10% of his total fighting force.[5] This number undoubtedly only increased as the Spanish sailed north around the British Isles. Though the Spanish did have 85 surgeons aboard the ships, these were doctors who were more adept at handling battlefield injuries than microscopic infections.

However, England was not exempt from the ravages of disease as well. Outbreaks of typhus and other illnesses struck the returning English sailors and encamped English soldiers. In July, a month before the Armada was driven back, Admiral Thomas Howard, 1st Earl of Nottingham, reported sick lists for his ships that grew by the week. More English souls were being swept away in the first weeks of the invasion by typhus, or gaol fever, than by the gunnery of the Spanish. By the end of the year over 7,000 Englishmen were in their graves as a result of the various pandemics that swept the British Isles. Howard memorialized the state of affairs in England in a letter to William Cecil in 1588:

> . . . It is a most pitiful sight to see, here at Margate, how the men, having no place to receive them into here, die in the streets. I am driven myself, of force, to come a-land, to see them bestowed in some lodging; and the best I can get is barns and outhouses. It would grieve any man's heart to see them that have served so valiantly to die so miserably.[6]

This was further exacerbated by bouts of food poisoning and associated toxemia that broke out among the great ships of the fleet. The cooking conditions and quality of water aboard naval vessels during the 16th century was undoubtedly more dangerous to crews than the fire of the enemy.

Likewise a return expedition by the English against mainland Spain suffered a similar fate to the Armada. In 1587, Sir Francis Drake sailed against Cadiz in an attempt to finish off the Spanish navy and perhaps begin

England's rise as a continental power again. Unfortunately for Drake, disease would carry off around 10,000 of his crew. This loss in personnel would cripple the English fleet and extinguish its hope of continental glory.[7] If England was to be able to compete with the likes of Spain and France, it would need to engage these powers in their own mercantilist and colonial game. Yet what would drive a sizeable population of England to leave their ancestral lands and proceed across the ocean?

The 1588 outbreak of epidemic typhus was only one of many similar outbreaks in England during the 16th century. Vectored by lice, the disease flourished in the atrocious sanitary conditions prevalent at the time. At the Black Assize of 1577 alone, 300 residents of Oxford died, including the local Baron. This particular outbreak was only the precursor to a typhus plague that swept England until 1579, killing an estimated 10% of the nation. Disease had become a common occurrence throughout Europe at the time, though this commonality did not reduce the fear it induced in the general populace. Plague outbreaks in London in both 1593 and 1603 killed a combined 34% of the city's population. Such epidemics resulted in societal strain, unemployment, and political turbulence; vital components for the creation of colonies and the mass exodus of peoples. Thanks in large part to the destruction of the Spanish Armada, the English now had both motives and opportunities to head west to the Americas and fulfill their desire to expand. Lice, more so than caravels, became the vehicles for the establishment of the English colonies.

CHAPTER THREE

POCAHONTAS AND THE PLAGUE
THE ENGLISH AND DISEASE IN
THE CONQUEST OF THE COLONIES

*"a strange mortalitie (sic) . . . affected a greater
part of his people . . . and but few escaped"*
—WEROANCE OF THE ACCOMAC, 1609

T he foundation of the American colonies was largely a messy affair. The
true history of English colonization lies behind the rosy tales of Pilgrim
Fathers, Thanksgivings, helpful natives, and religious and economic
freedom. The settlement of the Thirteen Colonies was a dirty, deadly, disease-filled endeavor. Nor was it a peaceful one for the English. From the first
settlement at Roanoke until the establishment of Georgia almost 150 years
later, wars and battles raged over the continent of North America. Once
again disease had a part to play in many of these encounters to conquer and
subdue what would become the Thirteen Colonies.

ROANOKE COLONY
Though the defeat of the Armada ushered in the era of English colonization,
its early history was one of disappointment and disaster. The initial settlements at Roanoke and Jamestown proved to be anything but a promising
start to what would eventually emerge as the famed Thirteen Colonies. Initially chartered by Sir Walter Raleigh, the colony of Roanoke was designed
to serve the economic interests of England. The arrival of Raleigh on July
4, 1584, was met with friendly natives and promising conditions. Positive
press soon followed, as exemplified by Richard Hakluyt, who wrote in 1584
of the Indians, *"the people good and of a gentle and amiable nature which willingly
will obey."* Hakluyt's words were prophetic, for soon after the arrival of the

first settlers the Indians did submit, not to the English, but to their accompanying diseases.

Soon after the arrival of Raleigh, the local Secotan tribe became the first group of natives to experience a decline in population that was becoming all too associated with European colonization. The Secotan wereoance (chief) Wingina had met and pleaded with Arthur Barlowe, an agent of Raleigh, for mutual cooperation against the Secotan's enemies. Though the talks amounted to little, the face-to-face contact between the two would have longer lasting results. Thomas Hariot would record how in the aftermath of the English journeys from town to town:

> but that within a few dayes after our departure from euerie such towne, the people began to die very fast, and many in short space; in some townes about twentie, in some fourtie, in some sixtie, & in one six score, which in trueth was very manie in respect of their numbers. This happened in no place that wee coulde learne but where wee had bene, where they vsed some practise against vs, and after such time; The disease also so strange, that they neither knew what it was, nor how to cure it; the like by report of the oldest men in the countrey neuer happened before, time out of minde. A thing specially obserued by vs as also by the naturall inhabitants themselves.[1]

The local Native Americans were baffled and Wingina and his council sought an explanation to the death being visited upon their people. Explanations ranged from the simple, with the English themselves being deliverers of the pestilence, to the improbable with the colonists' using *"invisible disease bullets,"* to the impossible, with the English being animated corpses. Perhaps most damning was the view of Ensenore, the chief advisor to Wingina, who put forth the proposition that the arrival of the English and the subsequent pestilence being visited upon the Algonquin peoples represented the wrath of the gods. Yet, as servants of the gods the English could not or should not be killed.[2] Though the wereoance was initially skeptical, the death of Ensenore to the same disease after he had spoken out against the colonists sealed the fate of the English. A plot was formed to ambush and destroy the small but threatening English colony. Though in the end the English preempted the Natives by seizing their leaders in a daring raid, a far greater damage had already been done. The settlers of Roanoke became a target, and combined with the failure of the English to resupply the colony during the Spanish War, the stage was set for the infamous abandonostracizedment of the colony before 1590.

The failed attempt by the English to colonize Roanoke in the late 16th century dampened the desire of those in the British Isles who sought to settle in or profit from the New World. The mysterious and dangerous land across the Atlantic was viewed as the home of whatever terrors the English mind could conceive. Yet the expedition did succeed in a far greater way than those back in London could have dreamed. Much as Cortez had laid the foundations for the destruction of the Aztec empire during his first failed visit to Tenochtitlan, so too did the English ensure future success in Virginia on top of the ruins of Roanoke.

The "Roanoke Disease," now tentatively identified as influenza, ravaged the natives of the region during the 1580s and 1590s.[3] By the time the English returned to the region in the first decade of the 1600s, they encountered a far different environment. The werowance of the Accomac tribe told John Smith during his June 1608 visit that, "*a strange mortalitie (sic) . . . affected a greater part of his people . . . and but few escaped.*"[4] The effects of the contagion were magnified by the fact that the region experienced at the same time the worst drought in centuries.[5] Crop failures from 1587 to 1589 would have weakened the local Natives' ability to fend off such diseases as influenza. Leadership of the region became insecure as the local chieftains succumbed to the plague. As well, the damage to the religious beliefs and

Settlement at Jamestown.

institutions of the Natives must have been profound as their traditional methods did little to stem the rising tide of deaths. Thus, though Roanoke failed as a colony, it was a necessary failure that allowed for the development of Jamestown.

CONQUEST AND WARS OF OLD VIRGINIA

Once the initial shock and fear associated with the Roanoke failure had subsided, the English began to plan for another major colonial attempt in 1606. Once again, economics was to serve as the driving force behind the undertaking. Thus in December of 1606 the Virginia Company of London dispatched an expedition of three ships and 144 males to the Chesapeake region of Virginia. No women were included on the voyage as the expedition was designed to be purely concerned with the mining of gold to send back to England. Not farming, nor exploration, nor permanent settlement were to be concerns of the colonists, only the short term goal of exploiting the mineral wealth of Virginia. In the end, the absence of this precious mineral at almost any point along the Eastern seaboard would prove to be the least of the colonists' problems.

While Jamestown would serve as the springboard for the colonization and development of America, its survival was highly questionable for the first two decades. The problems that buffeted the small settlement could have easily sent Jamestown the way of Roanoke and delayed if not doomed further English exploration. The tale of the colony is a well-known one. How the settlers battled the Natives, food shortages, greed, and each other to eventually emerge as a sustainable settlement. The troubles that beset the colony during its first few years in which a majority of the population died has historically been termed the Starving Time. Yet once again this is an oversimplification if not outright misnomer. What almost doomed Jamestown and Virginia was not a lack of food, but disease. While over its first winter a hundred colonists would starve, thousands would die over the next decade from a variety of both native and foreign pathogens. Disease was once again the factor that shaped history.

Despite the efforts of John Smith to save the colony by forcing every able bodied man to farm, the colony still faced many external as well as internal issues. John Rolfe famously married Pocahontas, with the union of the two serving to ease tensions between both cultures for a time. Yet during a trip by Pocahontas to England in 1617 the Powhatan princess would contract a disease, dying in John Rolfe's arms as they were departing London to return to America. The death of Pocahontas not only severed the links between the English settlers and the Powhatan, but rumors and diatribes against the English spread by Tomocomo, a Powhatan shaman, and others eventually led to a second, more disastrous war between the two groups. The war would begin with a well-orchestrated massacre by Chief Opechancanough aimed at 31 separate settlements and plantations. By the end of the Indian offensive, up to 25% of the colony had been killed off. In the end however, due to a combination of poor native strategy and the skillful and deceitful use of poison by the colonists, victory would eventually favor the Virginians. Disease brought about the final conflict between the English and the Natives which cleared the way for the further development of Virginia.

BACON'S REBELLION

The defeat of the coastal Chesapeake tribes allowed for generations of Englishmen to push further west, deeper into the New World. This migration would culminate in a minor episode in Virginia's history, the importance of which far outweighed its duration. By 1676 the colony of Virginia was expanding westward, a sure sign of its success. Yet the generations that had passed since the settlement of Jamestown had not produced a united region. Instead, Virginia was splitting into two very unique and divergent groups.

The lowlands along the coast contained some of the best land in the colony. Plantations owned by the wealthy controlled most of the landscape. These men tended to be of older stock, being members of some of the earliest families to colonize the region, thus of English blood and Anglican bent. Due to the geography of the region, the presence of malaria, and their level of wealth, their farms tended to be large plantations. This in turn necessitated large-scale slave labor by Africans or indentured servants. Their wealth, social status, and leisure time likewise resulted in their being made members of the House of Burgesses. Many, such as Governor Berkeley himself, carried on trade with the Natives in the colony, acquiring valuable furs in exchange for goods or weapons.

At the same time the upcountry was slowly emerging among the mountains and forests of central and western Virginia. Inhabited by later arrivals, most of these men were Scotch-Irish and Presbyterian. Many had received small parcels of land after serving their time as indentured servants. Their geography and lack of wealth combined to result in small subsistence-based farms, on which they eked out survival. Few if any could afford slaves, and they tended to be underrepresented in the House of Burgesses. Finally as these men pushed further west they came more and more into conflict with the Natives, many of whom fought with arms acquired from the landed English along the coast.

Thus it was only a matter of time before a physical split accompanied the socio-economic one that had already manifested within the colony. The catalyst for this was a series of violent Indian attacks along the frontier in 1675. Berkeley's failure to act (in large part due to a fear of losing the Indian fur trade) prompted the yeomen farmers to rally around Nathaniel Bacon in 1676. Employing a mixture of demagoguery, results, and alcohol, Bacon catapulted himself and the colonists' cause to the forefront and received a commission (after many threats of violence) from the House of Burgesses.

Nathaniel Bacon proceeded to attack and slaughter various Indian groups. Following this he turned his anger and that of the farmers against Jamestown and Berkeley. In July of 1676 Bacon penned the *Declaration of the People of Virginia*. Though not a call for independence, it was a striking polemic against the rule of Berkeley and the upper class of the colony. What followed was a civil war, in the course of which Bacon and his followers attacked Jamestown. After much fighting during rain-filled weeks, the rebels burned the colonial capital to the ground in September. The royal governor fled and sought for help from the King.

Bacon was on the verge of seizing the entire colony when, on October 26, 1676, he succumbed to dysentery and died. Though John Ingram and others took command of the movement, they lacked the flair and intelligence of Bacon. Both sides settled down into winter quarters, and by the spring of 1677 the rebellion was effectively quashed by Gov. Berkeley. The death of Bacon brought about the dénouement of the rebellion. Within months, what some have called the first American revolution had been consigned to history books.

The rebellion impacted Virginia in numerous ways. While some voting and civil rights were extended to the discontented colonists in order to prevent further civil disorder, the troublesome indentured servants became more and more undesirable as a source of labor. Virginia began to turn more towards racial slavery as a direct result of Bacon's Rebellion.

THE CONQUEST AND WARS OF NEW ENGLAND

In a similar fashion to the southern colonies of America, those in the north were likewise made possible due to the presence of disease. For years, fisherman from the British Isles had plied the waters off the northeastern coast of America. Yet, save for minor attempts at both Cuttyhunk and Popham to perfect settlements in 1602 and 1607 in search of sassafras (a reputed cure for syphilis), no major expedition was sent by the English north of Virginia until almost 1620.[6] As Virginia became more settled though, and colonization became more possible and commonplace, it became only a matter of time before these northern areas were targets of settlement as well.

On September 6, 1620, 102 English men and women boarded the *Mayflower* in an attempt to reach the New World. This was actually the second attempt by this group of religious Separatists—who called themselves Pilgrims—to depart from England, with the first sailing of two ships having to turn back after suffering from leaks aboard the *Speedwell*. While a more superstitious people would have taken this as an ominous portent, the Separatists were unwavering in their commitment to create a new, perfect society in America. Land was sighted two months later in November of 1620, but it was not the spot intended. Due to a combination of storms and design, the *Mayflower* arrived in the New World off the coast of Massachusetts, not the planned for mouth of the Hudson River.

Rather than proceed south to their proper destination so late in the year, the Pilgrims decided to land at what was to become Plymouth on December 17th. If the initial settlement at Jamestown proved anything, it was that landing in autumn or winter in hostile territory was deadly. Not only did the Pilgrims not learn from their predecessors a generation prior, but in fact faced a much tougher situation. The Separatists had arrived in a much colder region, much later in the year. The season was too advanced for the settlers to plant crops, and very few resources were found upon landing. As a harsh winter settled upon Cape Cod, the settlers were reduced to dire straits. Barely seven of the nineteen planned for structures were completed by the spring, and the colonists resorted to raiding local Indian graves for buried corn. By 1621, 45 of the 102 settlers had died, largely from scurvy and pneumonia; in addition tuberculosis was reported as rampant. Four of the families who landed in the Mayflower were completely wiped out to the last member, including the family of Christopher Martin, the expedition's treasurer and the leader of the Speedwell Pilgrims.[7] The Pilgrims lacked the medical materials and ability needed to counter the plagues besetting them. The closest thing to a doctor available was Deacon Samuel Fuller, a weaver and clergyman by training who may have attended a few lectures on medicine while the Separatists were in Leiden, Holland, prior to their departure to America.[8] By the spring of the next year only two individuals remained healthy enough to care for the remaining colonists, William Brewster and Myles Standish. Had it not been for the timely help of a local native named Squanto, the colony could have been lost. Even with this aid, if the encroaching Natives were to make a sudden attack on the colony, few doubted that Plymouth would not go the way of Roanoke.

Yet there was one clear difference between the colonies at Jamestown and Plymouth. After the Pilgrims made landfall, they noticed a distinct lack of Native hostility. More bizarrely they noticed a total lack of Natives whatsoever. A decade before the English landed at Plymouth, this region of New England was home to an estimated 144,000 inhabitants. The colonists upon landing, however, encountered more graves and bones than actual natives, with one colonist referring to the region as a new Golgotha.[9] To the Puritans this was surely the wrath of God, punishing the Natives for their heathen behavior. *"The natives, they are near all dead of the smallpox, so as the Lord has cleared our title to what we possess."*[10]

From around 1616 to 1619 the indigenous people of New England were visited by a pestilence that destroyed their society. It is estimated that much like in other parts of the Americas up to 90% of the Natives were exterminated. Only one tribe, the Narragansett, escaped relatively unharmed, with this most likely being due to their practice of burning their dead. The exact pathogen that visited the region has been much debated, with historians and medical experts suggesting smallpox, influenza, diphtheria, and other maladies. More recent evidence points to leptospirosis as a possible agent of the Natives' destruction.[11] As a disease that arises largely from the handling of animals, the hunter-gatherer nature of the Natives lent itself to such an epidemic. Regardless of the exact nature of the disease, its vector is better established.

As previously mentioned, English fishermen had plied the coast of New England for over a generation. In addition, at the time the Pilgrims landed they reported that they found graves that contained European remains. It is obvious that Europeans had a presence, albeit a small one in this region of New England for years before the Pilgrims arrived. As had been made evident repeatedly before this point, the presence of Europeans invariably spread disease through the Native communities. These fishermen and proto-colonists laid low the population of New England a decade before the arrival of the Separatists, thus paving the way for an easier colonization. The Puritans could rightly claim their success as an act of God.

The plague visited upon the natives of New England had an additional important effect upon the settlement of the region, producing a savior for the faltering colony. The story of Squanto is well known. How while the Pilgrims were at their most vulnerable, a Native American named Tisquantum appeared, a Native who both spoke English and was eager to aid the dying colonists. His actions ultimately helped to stabilize the colony. Squanto instructed the Pilgrims in Native planting techniques, and later served as a guide and translator for Edward Winslow and Stephen Hopkins as they attempted to make peace with the Wampanoag chief Massasoit.[12] In fact Squanto became so respected and loved by the Pilgrims that following his capture by the Wampanoag, a rescue mission was undertaken by Myles Standish to free him. Squanto continued to work with the Pilgrims until 1622, when he died after becoming feverish and bleeding from his nose, a condition termed by William Bradford as the *"Indian fever."*[13] It is undeniable that Squanto helped to secure the colonists' survival during their first two years in New England. Yet if it wasn't for disease, Squanto would have never ended up at the front gate of Plymouth.

As a young man Squanto was taken to England by George Weymouth, who in 1605 was exploring the coast of New England. Staying in England until 1614, Tisquantum sailed home with Thomas Hunt and John Smith on another exploring expedition. However, rather than be allowed to return to his village, he was sold by Hunt to the Spanish as a slave. Eventually freed by monks, Squanto joined yet another expedition to the New World, and after another failed trip across the ocean he eventually landed near his village in 1619. Yet this was far from a storybook ending. The village that Squanto had left behind in 1605 was found to be lying in ruins upon his return. Worse yet, the entire Patuxet nation had been destroyed. Disease had struck after Weymouth's expedition and had killed off almost the entire tribe.[14] Squanto was the last surviving members of an extinct people. When the Pilgrims landed two years later and established Plymouth town, they actually did so on the remains of an old Patuxet village. When Squanto wandered into the Pilgrims' world he did so as an exile. Had it not been for the epidemic visited upon his tribe, the region would not have been empty upon the arrival of the English settlers, and Squanto himself would not have been seeking out kindred human company. Perhaps the situation was best summed up by Robert Cushman in 1621:

> They were very much wasted of late, by reason of a great mortality that fell amongst them three years since; which, together with their own civil dissensions and bloody wars, hath so wasted them, as I think the twentieth person is scarce left alive; and those that are left, have their courage much abated, and their countenance dejected, and they seem as a people affrighted. And though when we first came into the country, were few, and many of us were sick, and many died by reason of the cold and wet, it being the depth of winter, and we having no houses nor shelter, yet when there was not six able persons among us, and that they came daily to us by hundreds, with their sachems and kings, and might in one hour have made a dispatch of us, yet such a fear was upon them, as that they never offered us the least injury in word or deed.[15]

The settlement of Plymouth, much like that of Jamestown, was not only burdened by disease but aided by it as well. Disease cleared the region of opposition to both English ventures, bound the colonists more closely together through necessity, and created the prospect of novel cultural, political, and societal evolvements to deal with the various contagions.

PEQUOT WAR

The New England colonies were just taking shape when a war broke out with one of the more powerful Native tribes of the region. For the previous century the Pequot Indians had been slowly moving down the Hudson River in an attempt to secure the region of Connecticut. Wars with the various tribes of the region raged for years interrupted only by the occasional epidemic and the arrival of the English at Plymouth.

The rise to power of the Pequot though was halted due to a number of factors. Foremost among these was the slow but steady expansion of the English inland and south along the New England coast. The Sachem Sassacus, only recently made leader of the Pequot, moved towards war with the European invaders. Yet only a year after he rose to power a terrible smallpox epidemic ravaged his people, carrying off vast numbers. The outbreak so severely weakened the Pequot that talk of war compelled a faction of the tribe to break away under Uncas and form the Mohegan who quickly allied with the English. The murder of colonist John Oldham proved to be the causus belli for open warfare, yet the demographic imbalance caused by the smallpox outbreak had already decided the war in favor of the colonists. The Mystic Massacre of a Pequot village that followed in 1637 doomed the cause of Sassacus and the war ended with the eradication of the Pequot as a nation.

THE CONNECTICUT WAR THAT WASN'T

The gradual movement west of the Puritans from Massachusetts was mirrored and opposed by the movement east of the Dutch from New Amsterdam. In the end the issue was solved not by military power but by epidemic strength. Two successive smallpox epidemics both propelled and preserved the English colonies along the Connecticut River and ensured that New England would stay English.

The ability of the Puritans to move into what would become the various colonies of Connecticut was brought about on the backs of *variola*. English immigrants to the New World had brought the smallpox virus to Massachusetts around 1630. Though only a handful of settlers are reported as dying from the outbreak, thousands of natives were killed or fled in terror from the pestilence. As more areas became depopulated of an aboriginal presence, the English were able to move inland and settle more regions.

Their success though sparked the ire of the Dutch, who were likewise moving into the Connecticut Valley region from New Amsterdam. The local Pequot Indians had encouraged the arrival of both sides in order to further their own interests in the fur trade. Yet after the English construction of what would become Newtown, the Dutch began to entreat the Indians to coordinate their efforts to drive the Puritans out. Hoping to cement an alliance, they sent a caravan to the Pequot loaded with trade goods, weapons, and foodstuffs. Unfortunately, it was also accompanied by smallpox. Infected individuals among the Dutch spread the illness to the natives. Gov. William Bradford of Plymouth reported that 95% of the native inhabitants of the Connecticut River died during the outbreak. Without support the Dutch were not able to drive out the English, and within only a decade dozens of new Puritan settlements dotted the landscape. Connecticut was founded as much by disease as by concentrated effort.

A similar smallpox outbreak in 1663 would significantly weaken the Dutch position at Ft. Orange, present day Albany. The arrival of an English invasion fleet the next year would encounter little resistance and seize the entire region between Massachusetts and Virginia with little effort. The Puritans remarked at the time that, *the hand of God had gone out against the people of New Netherlands by pestilential infections.*[16]

KING PHILIP'S WAR

The last great Indian war to be fought in the region of New England unfolded in 1675 with the launching of a desperate anti-English alliance. Headed by Metacomet, better known as King Philip, this war would become, in terms of percentages killed, the bloodiest war in North American history. Once again disease served as both an ally and adversary to the early settlers.

The origins of the conflict lay once again in the spreading of both the English and disease into the interiors of New England. Epidemics had erupted in the area among the Natives in 1616–19, 1633–34, 1647, and 1649–52. Dysentery, tuberculosis, influenza, and smallpox severely reduced the population of the Natick and associated tribes. Nor were the Puritans alone in viewing these outbreaks as divine retribution as it was also commonplace among many of the tribes.[17] Evidence points to a reduction in population from perhaps 90,000 in 1600 to around 11,000 in 1675.[18] It was a diseased, depressed, and depopulated confederation of tribes that finally engaged the English in King Philip's War.

Despite the losses and reverses faced by the English, pestilence proved to be a loyal ally. James the Painter, a converted Indian, told a colonist at the time that more Indians were dying from disease over the winter of

1675–76 than from bullets.[19] Initial successes by the Native tribes were eventually minimized by the vast reduction in population that they suffered. By the second year of the conflict the war was beginning to turn in favor of the English. As this happened, various other Native tribes began to likewise turn against Metacomet, undoubtedly seeking better terms once peace was achieved. Upwards of 70% of the region's Natives, friend and foe alike, perished in the war while the English lost perhaps 600–800 colonists.[20] King Philip's War permanently broke Indian resistance in the Northeast and assured that the English would remain.

Overall, disease was present at every moment of the initial English conquest of the Atlantic coast. From clearing out the Native population of the region and allowing for the successful Pilgrim landing, to reducing the power and effectiveness of the various Indian armies, to igniting a final war with the Powhatan, pestilence delivered the lands from New England to Virginia into the hands of the early American settlers. Great victories were able to be achieved with limited martial skill and power.

CHAPTER FOUR

"THE PATHS TO GLORY LEAD BUT TO THE GRAVE"
DISEASE IN THE EARLY FRENCH AND INDIAN WARS

"the Holy God sends diseases into our army"
—SIR WILLIAM PHIPS, 1690

The race to build colonies in the New World became another facet of the balance of power politics being played out across Europe during the 17th and 18th centuries. Therefore when war did erupt in Europe, it inevitably spilled over into the New World. The colonies became both battlefields and pawns for England, France, and Spain, and the death and destruction wrought by this did much to produce a growing wedge between the colonies and their colonizers. In the end, the Thirteen Colonies especially began to question whether or not their interests matched those of England. Were such episodes as sacrificing 18,000 men to yellow fever in an ill-conceived and failed attempt to take Cartagena, or losing 4,000 men to various tropical diseases in a siege of Porto Bello, or being required to quarter infected British troops in Philadelphia really in America's best interests? Disease would drive and shape many of these conflicts, and ultimately aid the English in dominating North America, yet at the same time it would also start the slow road towards American independence.

BEAVER WARS

One of the most profitable trade items in North America during the 17th century, and a driving force behind colonization, was fur. By the 16th century the forests of Europe were on the verge of depletion. The need for farmland, grazing pastures, firewood, and building materials had reduced

The Taking of Porto Bello,
by Samuel Scott (1741).

the amount of the continent covered by woodlands to only small areas deep within Slavic lands. At the same time, fur coats and felt hats were becoming fashion necessities in London and other major continental cities. Historically fur hats had always been a mark of bearing. Even Chaucer's merchant, to show his wealth and status was described as . . .

A Marchant was ther with a forked berd,
In mottelee, and hye on horse he sat;
Upon his heed a Flaundrish bever hat,
His bootes clasped faire and fetisly.

By the time of Shakespeare, the term "beaver" had become generically used for any lordly hat or helmet visor . . .

O, yes, my lord; he wore his beaver up. (Hamlet Act I, Scene 2)

Therefore as the prices of beaver hats and associated wares exploded due to the decreased supplies in Europe, the desire to find an alternative source became an impetus for penetrating the virgin forests of North America. Yet, as Europeans began to settle the seaboard and hunt for furs they quickly discerned the necessity of employing Indian labor. The local natives knew the terrain better, already had experience hunting the animals, and would work far more cheaply than would imported Europeans.

Two European powers initially battled for supremacy in the fur trade: the French and the Dutch. Coincidentally, two major Indian groups inhabited the fur region, the Algonquin and the Iroquois. These two groups had warred for centuries, with even French explorers such as Jacques Cartier in the 1530s recording engagements between the two. *"They gave us also to understande that those Agouionda doe continually warre one against another."*[1] The Dutch and the French began to arm their respective sides with European firearms. This served a dual purpose of allowing the Indians to both hunt more efficiently and to eliminate the other tribe more rapidly. The two major tribal confederations now began to battle each other in earnest to dominate the beaver market.

The balance began to tilt in the favor of the Iroquois Confederation during the 1630s due to the arrival of a new ally: disease. A series of pestilences descended upon the St. Lawrence region from 1634 to 1640, eventually becoming known as the Huron Indian Epidemics. From a population of almost 20,000 at the end of the 1620s the Huron nation was cut down to 9,000 souls by 1640. The origin of the diseases most likely lay in the holds of French ships making port at Quebec. From there, trade and contact between the French traders and/or Jesuits and their Algonquin allies would spread the illnesses. France's desire for trade became a self destroying force, eliminating the very tribes that provided the fur.

The first outbreak was reported around July of 1634 near Trois-Rivieres. Accounts from the time report a measles or smallpox- like epidemic accompanied by fever, rash, impaired vision or blindness, and diarrhea before the victim succumbed to death. The exact pathogen itself is debated, with many sources leaning towards a virulent form of measles. Due to the unfamiliarity of the Huron with the sickness, mortality rates were high.[2] Eighteen months later during the fall of 1636 an influenza outbreak raced through the region of southern Ontario, further decimating the Algonquin tribes. Just as the influenza was subsiding in the summer of 1637 a new illness, possibly scarlet fever, exploded in the region as well. Finally in 1639 a smallpox epidemic attacked what few, weakened natives were left. *"Terror was universal. The contagion increased as autumn advanced and when winter came, far from ceasing, its ravages were appalling."*[3] Many of the natives blamed the traveling Jesuits and their baptismal practices, damaging the hold and influence of the order on the Indians of the region.

Less than a year after the Huron Epidemic had ended, war resumed between the Algonquin and the Iroquois. The former was so weakened by the various epidemics of 1634–40 that they proved unable to hold back this latest assault from their traditional enemies. The security of French trade in the region and the concern of the Jesuits for their conversion efforts combined to convince the French authorities to push for peace. By 1645 the French had brokered a peace deal with Deganaweida and Koiseaton of the Iroquois Confederation, granting most of their demands including the ability to trade furs in Quebec.

The peace was short-lived. By 1646, the French were refusing to trade with the Iroquois and war erupted again. Desperate, the Huron allied with the Susquehannock, a traditional enemy of both the English and the Iroquois. However, these efforts amounted to little. By 1649, the Huron were defeated. The victory of the Iroquois was owed as much to the guns of the Dutch as it was to their diseases. As one French allied native informed a local Jesuit priest, *"You tell us that God is full of goodness, and then, when we give ourselves to him, he massacres us. The Iroquois do not believe in God, they are more wicked than demons, and yet they prosper. . . . Why did he not begin with the Iroquois?"*[4]

The effects of the Beaver Wars were both numerous and decisive. To replace their own significant losses the Iroquois began adopting Huron into their nation. A letter from Jacques Bruyas reports that by 1668 almost two-thirds of an Oneida village he had visited consisted of Huron rather than of Oneidas.[5] Yet the Iroquois themselves did not escape the ravages of pestilence. Smallpox erupted among the Onondaga in 1656 and among the other tribes by 1661. The Seneca were visited by smallpox in 1668 and then by an outbreak of influenza in 1676. Soon the Iroquois were little better off than the Huron. From 1640 to the 1670s the Mohawk population alone was reduced by half, with its available number of warriors reduced from 800 to 300.[6]

Yet war, death, and loss were nothing new to the Iroquois. As with most small, clan-based societies, population size was a perennially vital concern. Thus the various tribes had created a solution to both the issues of emotional loss and population size generations before. A family, village, or tribe that had recently suffered a loss could launch a "mourning war" against a traditional enemy (not necessarily a guilty party) as a means of restoring depleted population or compensating for death. These raids or wars, depending upon their size, could be used to gain scalps, goods, or captives. If prisoners were taken they could be subjected to adoption, torture, or execution, depending upon the whim of the village and the family of those in mourning.

> Subsequently addressed appropriately as "uncle" or "nephew," but his status was marked by a distinctive red and black pattern of facial paint. During the next few days the doomed man gave his death feast, where his executioners saluted him and allowed him to recite his war honors. On the appointed day he was tied with a short rope to a stake, and villagers of both sexes and all ages took turns wielding firebrands and various red-hot objects to burn him systematically from the feet up. The tormentors behaved with religious solemnity and spoke in symbolic language of "caressing" their adopted relative with their firebrands. The victim was expected to endure his sufferings stoically and even to encourage his torturers, but this seems to have been ideal rather than typical behavior. If he too quickly began to swoon, his ordeal briefly ceased and he received food and drink and time to recover somewhat before the burning resumed. At length, before he expired, someone scalped him, another threw hot sand on his exposed skull, and finally a warrior dispatched him with a knife to the chest or a hatchet to the neck. Then the victim's flesh was stripped from his bones and thrown into cooking kettles, and the whole village feasted on his remains."[7]

Despite the defeat and adoption of the various Huron tribes by the Iroquois, disease continued to ravage and deplete their population. In 1701 a smallpox outbreak erupted in the Catholic Mohawk village of Kahnawake, resulting in many deaths. In response, a contingent of Mohawk joined the French and Abenaki in the infamous Deerfield Massacre as part of a mourning war. Overall, 56 residents of Deerfield, Massachusetts would be killed and 112 made prisoners by the various tribes who conducted the raid. As the various French and Indian Wars erupted over the course of the next half-century, mourning raids by the various tribes became more and more common. Though to the French these attacks were strategic assaults on English frontier villages, to the Natives they became more and more desperate attempts to relieve sagging village populations brought about by the invisible enemy of disease.[8]

An additional result of the Beaver Wars was its impact on French power in the region. With their trading ally removed and the beaver largely depleted the French had to push deeper into Canada to acquire furs. As well, Quebec's southern frontier was now left open to Iroquois and British incursions. Direct assaults on French held posts became more frequent and more damaging. In 1687 Iroquois war parties attacked Forts Niagara and Frontenac in response to a raid by Marquis de Denonville against the Seneca along the Genesee River. Having attacked up and down the river, de Denonville erected a stockade at the future site of Ft. Niagara in July. Leaving Capt. Pierre de Troyes in charge, he withdrew with the larger part of his force towards Quebec, just as the Iroquois began to besiege the bastion. Food and supplies soon ran short and disease erupted among the garrison. By the time the river was navigable in the spring of 1688 and reinforcements arrived, only 12 of the 101 French soldiers in the fort remained. The scenario repeated itself in September, and it would take over three decades for the French to finally establish a presence in the region. Father Millet, part of the original relieving force sent in the spring of 1688, erected a memorial cross, a copy of which still stands today, dedicated to the memory of those who succumbed in the fort to the various plagues.

A similar situation unfolded at Fort Frontenac in 1688. Scurvy soon overtook the garrison much as it did at Ft. Denonville. By 1689 the French had abandoned the base along with 180 of the 240 men stationed there. The war was turning against them as various towns were massacred and the natives threatened Montreal itself. De Denonville was removed from command, and as disease and war were crippling both sides, peace was eventually achieved in the famous Great Peace of Montreal in 1701.

KING WILLIAM'S WAR

As European warfare became more endemic and more destructive, the various great powers sought to isolate their colonies so as to preserve an economy of force in the European theater. Therefore in November of 1686 the leaders of England and France met and signed the Treaty of Whitehall, declaring the North American colonies to be neutral in case of conflict between the two nations. Within two years the League of Augsburg, including England, the Netherlands, Spain, and various other powers, was at war with the French kingdom of Louis XIV. Despite the hopes expressed at

Whitehall, by the middle of 1688 the war had spilled over into the North American colonies. Known as King William's War in the Americas, this decade-long conflict produced nothing for the English and French colonists but death and disease.

Though the population of the English colonies in America outnumbered that of New France by almost 15:1, the war began early to turn against the English. The French had seized all three English forts along James Bay back in 1686, despite the fact that the two nations were still technically at peace at the time. In 1688 the Hudson Bay Company dispatched several ships to reestablish control over Ft. Albany, Rupert House, and the York Factory. Arriving in September aboard the *Churchill* and *Yonge*, English soldiers began to construct simple siege works outside Ft. Albany. With 85 men in their landing force against only 16 French Canadians and some sailors inside the fort, the English felt confident in victory. Yet this enthusiasm was short-lived.

With the onset of winter in November, the English ships became frozen in the bay. The French bided their time, knowing that each day that passed would result in continued attrition of the enemy. Scurvy began to erupt among the soldiers, quickly reducing the force to a shadow of its former self. To augment the strain upon the British, Pierre Le Moyne d'Iberville, the French commander, both denied the enemy a ceasefire in order for them to hunt for game and at the same time captured the only English doctor in the force in a bold coup.[9] The besiegers soon became the besieged, and growing desperate for food and warmth dispatched a large woodcutting party to secure fuel. D'Iberville fell upon these men in the wilderness, taking 20 of them captive. After this the French commander felt confident enough to assault the barricades around his fort. When he did so he quickly discovered that disease had so reduced the besieging force that only eight Englishmen remained fit enough to confront the French. The English invasion force and their two ships fell to d'Iberville. Scurvy and the French assault upon the medical service of the English had prevented what should have been a sure English victory. It would be five years before London would again control Fort Albany. D'Iberville himself would go on to assault and

Pierre Le Moyne d'Iberville

take York Factory along the Bay in 1694. However, as the outpost fell so late in the year, he was forced to winter in it with his English prisoners. Scurvy reduced both sides to such an extent that the English were able to retake the post the next year.

With the onset of hostilities came the raids that had characterized colonial warfare for generations. In August of 1689 the Iroquois raided the French settlement of Lachine near Montreal, while the next year the French under d'Iberville launched a raid against Schenectady. According to local lore the town's stockade was open and the only sentinels were two snowmen silently guarding the entrance. Within hours, 60 villagers of all ages and sexes had been killed and 27 had been taken prisoner in mourning war fashion. The importance of these episodes is the resulting call among the colonists and British for an assault on Quebec to permanently reduce the French colony and thus safeguard settlements in the New England region.

Sir William Phips was appointed by the General Court of Massachusetts with the task of leading the expedition against Canada. He had previously received praise for his assault and capture of Port Royal on May 21, 1690. The French commander of that town, de Menneval, had few men and no mounted guns and was unable to provide any significant resistance. Upon entering the town Phips' men proceeded to wreck and plunder the local churches and domiciles. He returned to Boston a hero and the obvious pick for a subsequent assault on Quebec. The English plan called for a two-pronged move against Quebec while a holding force took Montreal and prevented reinforcement of the capital.

The first arm of Phips' pincer was dispatched from Hartford on July 14 of 1690 under Fitz-John Winthrop, the grandson of the more famous John Winthrop. The army of over 600 men from New York, Connecticut, and local allied tribes advanced up the Hudson River in a bid to launch a coordinated attack on Quebec that fall. Yet disease soon began to appear among the eager colonial soldiers.

> I must allsoe mention to you with great sorrow that your army is much disabled with sickness; the small pox, the feavor & the flux is very mortall, 4 or 5 haueing dyed in these few dayes of my being here, and every day more are visited, and vpon a veiu of the severall companyes I cannot depend vpon aboue 130 soldiers fit for service,

and every ones apprehention of being taken with the distemper vpon the march, soe remote from all help, does wholy discourage them, and in truth it is very reasonable to consider about it, since without a miracle of Providence none can escape under such a visitation.[10]

By the time the force had reached Lake George, a distance of less than 170 miles, smallpox had crippled the invasion.

Many, haueing been taken sick on the march, are returned, and since my being here seuerall are taken ill, and Liuetenant Hubble and an other downe of the small poxe, which, thus remote in the wilderness, is double affliction. Under these misfortunes . . . I am surrounded with difficultyes, and ouer prest in my minde to finde your designes obstructed . . . wee must submit, remembering that not one hayre of our heads falls to the ground without Gods appointments.[11]

The promised Indian allies had likewise been reduced from hundreds to dozens as, *"the great god had stopt (their) way,"* through the outbreak of disease.[12] The expedition collapsed thereafter with Winthrop being arrested for treason, having to petition the king for redress over the course of the next decade.

After his removal to Albany, Winthrop dispatched Captain John Schuyler with 150 Albany militiamen and Indians to launch a smaller, guerrilla style assault deep into the wilderness of Canada around Montreal. Lack of supplies combined with smallpox to reduce Schuyler's effective force to a fraction of both what was promised and what was needed to pin down French forces at Montreal. His ragged force emerged from the woods south of Montreal in early September, and after killing 50 villagers returned to New England. The failure of this expedition assured a lifeline for Quebec and allowed 200–300 additional French fighters to reach the capital before the arrival of Phips.

The invasion of Quebec itself was left to the main force of Sir William Phips who sailed with high hopes as word of Winthrop's failure did not reach Boston until a week after the invasion fleet had already left, and

Schuyler's force was still incommunicado. The general had been quite leisurely about his departure, waiting until late in the summer for the arrival of more supplies from England. When the invasion force of 32 ships and 2,300 men did depart from Hull on August 9th Phips continued his leisurely pace, not arriving at Quebec until October 16th. Thanks to the defeat of the Schulyer Expedition, the capital had been reinforced only two days prior. Frontenac was secure in his newly fortified city with around 3,000 men at his disposal.

Much worse for Phips, smallpox had erupted on the decks of the invasion fleet as it tacked out of Boston harbor. The ship's surgeon, Dr. Edwards, died on Oct. 4th, a forewarning of more losses to come. It was to be a reduced and weary force that assaulted Quebec in mid-October. Phips reports that only 1,400 men under Major Walley were well enough to land at Quebec on October 7th.[13] By the 23rd the battle was over with Quebec remaining in French hands. While only 30 New Englanders succumbed to death in battle, over 1,000 had died from smallpox. Phips returned to Boston on November 19th, repeatedly blaming the failure of Winthrop, his Indian allies, and smallpox for his inability to take Quebec in various accounts written by himself.

> Not ye enemy but Almighty God himself did frustrate our design. His finger is to be seen in visiting ye Indians with ye Small pox . . . the Holy God send diseases (a malignant fever and ye Small pox) into our army.[14]

Again . . .

> at the same time fourteen hundred men that I had landed defeated a great part of the enemy, and by the account of the prisoners, the city in all probability must have been taken in two or three dales, but the small pox and the feavour increased so fast among the men that it delayed the pushing on the siege till the weather grew so extream cold that no further progress could be made therein.[15]

The failure of Phips' Expedition reverberated throughout the colonies. Henry Sloughter, the governor of New York, remarked, "*the whole country from Pemaquid to Delaware is extremely hurt by the late ill managed and fruitless expedition to Canada, which hath contracted £40,000 debt and about 1,000 men lost by sickness and shipwrack and no blow struck for want of courage and conduct in the Officers."* This debt would weigh heavily upon New England as it was issued in paper by the English. Phips did much to attempt to alleviate this debt by buying it up, eventually crippling himself financially. In addition, the attack prompted Frontenac and the French to further augment the defenses of Quebec, which were completed by 1693.

Further raids would take place with the French and their Abenaki and Algonquin allies ravaging New Hampshire and Maine. Hundreds more would die, but the war for all intents and purposes had reached a stalemate by the 1690s. A final offensive by the British navy against the Windward Islands in 1693 did little more than create an outbreak of yellow fever which resulted in the death of over 3,100 men out of a force of 4,500. When the sickened and dying fleet under Sir Francis Wheeler docked at Boston, it ended once and for all English hopes of taking Quebec. As recorded by Mathers in his *Magnalia Christi Americana*:

> There was an English fleet of our good friends with a direful plague aboard 'em, intending hither. Had they come, as they intended, what an horrible desolation had cut us off, let the desolate places that some of you have seen in the colonies of the south declare unto us; and that they did not come, it was the signal 'hand of Heaven', by which the 'goings of men are ordered.[16]

With the signing of the Treaty of Ryswick in 1697, the status quo was returned to North America. Despite the sacrifices of the colonists, including thousands who succumbed to smallpox and scurvy, nothing was gained. Worse yet, the Iroquois, as previously mentioned weakened by disease and abandoned by their ally, sought permanent peace with France at Montreal in 1701.

QUEEN ANNE'S WAR

Only five years after the formal ending of King William's War and a year after the Great Peace of Montreal, war again erupted among the great powers. Beginning in Europe in 1701 as the War of Spanish Succession, the conflict soon spilled over into North America much as had the War of the Grand Alliance a generation prior. Raids against Newfoundland and New England by the English and the French, along with their respective Indian allies again dominated the northern theater of the war. However with England's declaration of war against Spain in 1702 the southern colonies suddenly became more actively involved in the colonial wars that had plagued New England during the previous century.

Disease once again plays an interesting role in a variety of episodes during the conflict, especially in the southern theater. Four years into the war the French and Spanish began to hear rumors of a yellow fever outbreak in Charles Town, South Carolina. As Georgia had not yet been established as a buffer colony, Charles Town, a city founded only 36 years earlier, was all that stood in the way of a Franco-Spanish invasion of North Carolina and Virginia. Steady immigration had made the settlement the fifth largest city in the Thirteen Colonies by 1690. Yet it was anything but a serene destination for immigrants and settlers. In 1699 alone the city had been visited by a fire which had destroyed one-third of it, then a hurricane in the autumn, a smallpox epidemic that had claimed the lives of almost 300, a yellow fever outbreak that killed 160, and an earthquake. In total the year saw 15% of the city's population buried. The Franco-Spanish invasion of 1706 must have been seen as the coup de grace for a doomed colony.

Organized by the hero of the Hudson Bay campaign, d'Iberville, a French invasion fleet of 12 ships and 600 soldiers departed Europe in January of 1706. Having stopped at Martinique for further reinforcement and having ravaged the English island of Nevis, the squadron called at Cuba to raise Spanish forces for an attack on the southern colonies. Yet a yellow fever outbreak was at that time crippling Havana, and few troops were made available. Worse yet, the disease soon began to sicken and kill the French soldiers, including the great d'Iberville, who would die in early July. By the time the new commander, Lefebvre, departed for Florida only 300 French and 200 Spanish soldiers remained. Picking up a further 30 Spanish and 50 Indian infantry at St. Augustine, the French and Spanish invasion force was launched on August 28, 1706, upon receipt of information regarding the epidemic then sweeping Charles Town.

While the French and Spanish force may have been heavily reduced by disease, losing both soldiers and an irreplaceable leader, they still expected an easy victory against the plague-swept city. Calls for Governor Nathan Johnson to surrender, however, proved fruitless. Worse yet, despite the presence of yellow fever which carried off 5% of the city's population, its defenses proved stronger than expected. Having heard rumors of the impending invasion, hundreds of militia had gathered from around the Carolinas to defend the capital. The governor had wisely stationed these men a mile and a half outside of the city to avoid exposing them to the contagion. When the Franco-Spanish assault finally took place on September 2nd, it was met with determined colonial resistance. Over 300 Frenchmen and Spaniards would be killed or captured while less than a handful of South Carolinians would fall. Yellow fever had both crippled the Franco-Spanish invasion force and lured them in to a foolhardy assault on Charles Town. At the same time, Gov. Johnson's strategic precautions during the epidemic saved his city from destruction.

The various wars of the period also produced a novel character of the age, the privateer. Licensed pirates, these men plied the Caribbean and other waters waging war for personal gain against the enemies of their nations. One such product of Queen Anne's War was Edward Teach, better known as Blackbeard. He had operated out of Jamaica during the war, before eventually transforming into a full pirate like so many other privateers of the period. His ship *Queen Anne's Revenge* began to attack various European ships up and down the Atlantic coast from 1716 until 1718, with Teach becoming legendary due to his bravado and theatrics. By 1718 Blackbeard even became bold enough, or perhaps desperate enough, to raid Charles Town. Only 12 years after the city was saved from a large Franco-Spanish invasion, Blackbeard held the entire community for hostage, capturing its leading citizens with only his small flotilla. Yet the pirate king was not after gold or gems; instead he demanded from the town leaders in exchange for their lives and the survival of their city, medicine. Historians have speculated for years that Teach and perhaps many members of his crew were suffering from venereal diseases, requiring the mercury and other chemical agents that would be present in the medical chests available in Charles Town. When his ship was discovered and excavated centuries later, urethra syringes containing mercury were found in the wreckage, providing further evidence for the theory.

Teach's attack on Charles Town would be his last major raid. In June of 1718, he and most of his crew accepted a pardon from Governor Eden, settling down as private citizens in the Carolinas. Yet the siren call of the ocean proved to be too much and Blackbeard was soon again terrorizing the coastline. Memories of his various raids, especially that on Charles Town, died hard, and the English soon ordered his arrest and execution. Venereal disease had lured the famed pirate into his largest and last battle.

WAR OF JENKINS' EAR

Relations between Spain and Great Britain remained strained following Queen Anne's War. Disputes over the slave trade soon led to increased Spanish seizures of British merchantmen. As a result of this, anti-Spanish sentiment slowly grew in England until it exploded into a popular demand for war following the assault on Robert Jenkins by a Spanish captain in which, having been accused of smuggling, Jenkins had his ear cut off by the Spaniard.

For the next nine years, from 1739 until 1748, the British and Spanish engaged each other around the world, with some of the harshest and most determined fighting taking place in the Caribbean theater. The potential economic windfall to be gained from control of the West Indies and the Latin American coastline lured the English into a series of doomed attacks. As always, it was not a lack of tenacity or skill on the part of the British that defeated their efforts, but the age-old nemesis of disease.

The draw of Latin America had previously enticed the British in 1726 to attempt to seize Porto Bello, Panama. It was hopped that a quick and decisive raid upon the fortress would enable the British to prevent the annual Spanish treasure ships from reaching Europe, thus crippling the Spanish economy and prevent them from possibly allying with Austria. In March of 1726 Rear Admiral Francis Hosier departed England for the Americas. His armada consisted of around 20 ships and 4,750 men. Reaching Panama in June, Hosier proceeded to blockade Porto Bello as ordered by Prime Minister Walpole. Yet by December, yellow fever had become so rampant among the crew that Hosier was forced to give up his blockade and withdraw to Jamaica to rebuild his force. During his absence the Spanish treasure fleet was able to sneak past and depart for Spain with close to 31 million pesos in its hold. Between 3,000 and 4,000 Englishmen had died to achieve nothing. By August of the next year Hosier himself had succumbed, and his replacement, Edward St. Lo, suffered a similar fate in

Admiral Vernon, by
Thomas Gainsborough

1729. In the end the entire undertaking proved to be a disaster. By November of 1729 England agreed to conclude the campaign with the Treaty of Seville, achieving little for their great sacrifice.

With the opening of the War of Jenkins' Ear, the English again aimed at launching an attack on Porto Bello, Panama. The architect behind this second expedition was Vice Admiral Edward Vernon. Having studied the previous assault of a decade prior, Vernon argued for using a small squadron to quickly assault and take the fort, rather than employing a larger fleet to blockade it. It was hoped that this would prevent the outbreak of disease that had so crippled Hosier's efforts. The expedition departed from Great Britain in July with only six ships and two tenders. After a quick assault, Porto Bello fell to the English on November 22, with the British suffering only three men killed. For three weeks, Vernon's men ravaged the city and region, severely damaging the economy of both Spain and Panama. The victory did much to erase the shame of the first Porto Bello raid and showed that with proper leadership, tactics, and preparation, disease could be prevented or at least mitigated.[17]

As a wave of patriotism swept England, another expedition was launched to further disrupt Spanish shipping. Commodore George Anson was dispatched in 1740 with six warships mounting 232 guns on a voyage to round Cape Horn and seize Lima, Peru. From the beginning the expedition was beset by troubles. The 500 troops proposed for the voyage were raised from 500 invalids from Chelsea Hospital. Worse yet, those who could avoid service did, with only around 259 showing up for duty, many of whom were carried aboard on stretchers.[18] Two months into the voyage there were already concerns over the rotting of food and the large number of flies inhabiting the ships. The vast amount of supplies that needed to be loaded aboard the ships caused its gun ports to be below the waterline. Anson ordered air holes to be cut into the sides of the ships in order to provide fresh air for the crews.

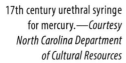

17th century urethral syringe for mercury.—*Courtesy North Carolina Department of Cultural Resources*

Once in the Atlantic, typhus and dysentery, the traditional sicknesses of the sea, began to ravage the crews. By December the fleet had crossed the Atlantic, calling at the Portuguese island of Isla de Santa Catarina to unload the sick and clean the ships. Eighty men had to be removed from the *Centurion* alone to recuperate on land. Fire and vinegar were then applied to the decks and holds of the warships to battle the fleas and pestilence which were decimating the crew. "*These operations were extremely necessary for correcting the noisome stench on board, and destroying the vermin; for from the number of our men, and the heat of the climate, both these nusances had increased upon us to a very loathsome degree; and besides being most intolerably offensive, they were doubtless in some sort productive of the sickness we had laboured under for a considerable time before our arrival at this island.*"[19] Though this was accomplished, the men ashore soon began contracting malaria, adding to the misery of the fleet. Twenty-eight men from the *Centurion* would never leave the island, while the majority of ships ended up taking back aboard more sick then they had initially disembarked. Finally, in March of 1741 as the ships rounded the tip of South America, Anson reported scurvy rampant among his men, resulting in hundreds of more cases of illness and death.

Conditions grew terrible and the ships became separated. The *Wager*, with too few men left healthy enough to main her sails, wrecked off the coast of Chile. The *Centurion* sailed for the rendezvous point at Juan Fernandez Island, and lacking men fit enough to lift the anchor, had to wait for a sudden squall to blow them into the harbor. The *Tryal*, which had lost 46 of 86 men, had only the captain and four sailors left in any condition to pilot her into the harbor next to the *Centurion*. These two warships were soon joined by the *Gloucester* and the *Anna*, though the latter was found to be too damaged to continue and was soon broken up. Meanwhile, the *Severn* and the *Pearl*, having also suffered damage and loss while attempting to round the Horn, decided to return home to England. The *Severn* herself reported 158 dead and 114 sick, with just 30 sailors left healthy enough to man the ship.

After taking time to refresh his men and resupply his ships, Anson prepared to continue the voyage. Before departing he took a census of the fleet, finding that of the 961 men who had departed England aboard the various warships, about 2/3 of that number, or 626 men, had died. Though his force had been severely reduced in strength, it fared better than the Spanish fleet under Don Joseph Pizarro that had been shadowing it since leaving European waters. Disease and storms had wracked the Spanish fleet, eventually forcing it to give up its pursuit. Anson turned his ships north to prowl the South American coast. Yet conditions aboard his warships were still not ideal, and at one point the *Gloucester* had to employ Spanish prisoners to man her sails.

After successfully raiding Spanish shipping as far north as Mexico, Anson decided to head back to England by the perceived safer route of crossing the Pacific. However due to unfavorable winds the crossing of this ocean became as deadly as the crossing of the Atlantic. Disease and scurvy quickly reappeared, decimating what was left of the crew. By August, only the *Centurion* with around 250 men aboard remained afloat, but with upwards of ten men dying every day, its fate seemed sealed. Anson landed on the Spanish held island of Tinian, northeast of Guam on August 23rd. Of the 128 sick put ashore, 21 would die during or immediately following the landing. Native breadfruit helped to revive the health of the men and repairs were undertaken on the slowly sinking ship. The *Centurion* soon set sail again, finally reaching the relative safety of Macao on November 11th.

Following more repairs Anson set out to turn his to-date lackluster campaign into a success by capturing the Spanish treasure ship sailing from Acapulco to Manila. Upon accomplishing this and returning to England, he was proclaimed a hero. Even the ordinary seamen of the fleet were richly rewarded, receiving the equivalency of 20 years' wages. Yet as has been seen, the cost in human life and misery was excessive. Only 188 men from the *Centurion, Gloucester, Tryal,* and *Anna* survived. Including the ships that returned to England after having failed to round the Cape, only about 25% of the original force made it home alive. Scurvy, typhus, malaria, and dysentery had destroyed almost 1,400 men and four of his majesty's ships.

Yet the one important product of the expedition was the knowledge it imparted to the English concerning disease at sea. Dr. James Lind would contribute through his research more to the English navy and people than did Comm. Anson through his voyage. Following the return of Anson, Lind began to perform clinical trials upon various cures for scurvy. By 1753 he penned a treatise correctly identifying citrus fruits as a cure and preventive measure for the illness. Though Lind was by no means the first to discern this link, his writings can be said to have been the ones that popularized the concept of a treatment. By 1758 he had been made chief physician of the Royal Hospital at Haslar, and a few years later based upon his observations at that institution wrote "An Essay on the Most Effective Means of Preserving the Health of Seamen" (1762).

Many of Lind's recommendations were eventually adopted by the British military, and by diffusion the American as well. These included the practice of cleaning and shaving soldiers and providing them with fresh linens as a way to combat typhus, a connection that he drew both from his time at Haslar as well as from anecdotal evidence. In a letter written in 1793, Lind describes India as being able to avoid typhus due to its residents' *"daily ablutions in their sacred tanks and rivers. . . . Their being either altogether without clothing, or using only those of muslin, which are washed every day."*[20] The doctor likewise authored papers and popularized ideas to deal with the issue of scurvy at sea, advocating the drinking of fresh steam-distilled water rather than beer and the procurement of fresh fruits.[21]

Yet Lind's ideas were too late for the final English campaigns of the war. As Anson's fleet was rounding Cape Horn, the British were dispatching a second fleet to once again invade and ravage the Spanish Main. Based upon their successful raid on Porto Bello, London aimed for a larger prize by taking the main Spanish ports in the region including Cartagena.[22] The important port city of around 10,000 had previously been assaulted by both Drake in 1585 and Baron de Pointis in 1697. On paper it was a heavily defended location, surrounded by a moat and numerous walls and bastions that mounted around 160 guns, with another 140 within the outskirts of the town itself. In addition, with little to no fresh water available outside of the city, a besieging army would be at a severe disadvantage during a prolonged campaign.

Dr. James Lind

The expedition was organized under the command of Adm. Edward Vernon, the hero of the lightning raid on Porto Bello two years previously. The English departed for the Caribbean with 27,000 men and 186 ships. The massive fleet mounting 2,000 guns and containing two hospital ships, 80 troop transports, and over 12,000 soldiers was thought to present an assured victory if the small fleet that took Porto Bello served as any model. The local Spanish leader Don Blas de Lezo had available only 3,000 soldiers and perhaps 5,000 natives. Yet the very success of Porto Bello resulted from the small size of the English contingent, its rapid deployment, and the decisive strike that it made upon the fortress— the opposites of what the English intended in 1740.

Vernon was slow to leave England for a variety of reasons. As a consequence, a major typhus epidemic then sweeping the nation proceeded to wreck havoc on the fleet. Over 25,000 men in the British navy fell ill, 2,750 would die, and almost 2,000 more would desert. In the meantime, the Virginia colonial government under Sir William Gooch had raised the 43rd Regiment of Foot to serve alongside the British at Cartagena. One of the four captains appointed by the governor was a young Virginian named Lawrence Washington, the 23-year-old elder brother of America's future founding father. The American force was eager for action and by December of 1740 had already arrived in Jamaica, well ahead of the British fleet. Yet a lack of provisions quickly took its toll on the expeditionary force, as did the various tropical fevers of the island. By January of 1741 both the American and English forces were experiencing growing sick and casualty lists. Over 500 men had died and 1,500 were too sick to muster. Included among the casualties was the commander of the expedition, Lord Cathcart, and his American counterpart, Lt. Gov. Alexander Spotswood.[23] Cathcart was subsequently replaced by Thomas Wentworth, a man with little previous command experience.

In early March, the British fleet finally arrived at Cartagena, having left a trail of burials at sea across the Caribbean. Sickness had so reduced the armada's strength that a third of the soldiers were mustered to fill the positions left open by sick or dead sailors.[24] Yet, initial assaults were promising. The Boca Chica forts protecting the harbor were soon cleared at a loss of only 120 killed and wounded. Though during the same period of time as the assault 250 more Americans and English would die from malaria and yellow fever, and 600 more would fall ill. In fact as the "sickly season" began in May of 1741, disease reduced Wentworth's available force from 6,500 to only 3,200. Additional assaults on various bastions began to grind to a halt, and by May 7th only 1,700 men were deemed fit for duty. Vernon and Wentworth eventually agreed to withdraw from battle, boarded their transports and returned to Jamaica. Yet the pestilence among the army could not be fled from. Within a month after departing another 1,100 were dead. The force that returned to safety in the British islands of the Caribbean consisted of less than one-third of the men who had originally sailed for Cartagena.

In all, the expedition lasted for 67 days and saw the deaths of 18,000 men, mostly from disease. Only 300 Americans would return to Virginia, including young Lawrence Washington.[25] King George II was allegedly quite distraught, forbidding anyone from talking or writing about the failed expedition. Medals previously struck to commemorate Vernon's promised easy victory were quietly and quickly removed from public sight. The government of Walpole collapsed shortly afterward, as did Britain's support of the Pragmatic Sanction, thus helping to stir up the War of Austrian Succession. Yet the Spanish themselves did not escape unharmed from Vernon's failure to take Cartagena. Not only did the pestilence that beset the English also visit the Spanish garrison, but the number of dead left behind by the retreating army caused further disease and havoc. De Lezo, the Spanish commander, would himself die a few weeks after his victory due to an illness contracted in the process of removing the English and Spanish dead.

Meanwhile, Adm. Vernon wagered on one last assault in the Caribbean. Seizing Guantanamo Bay at the beginning of the summer of 1741, Vernon then proceeded to attack Santiago de Cuba from July until December. His expedition of 61 ships and 4,000 men camped outside the city for four months. Yet once again tropical and camp diseases spelled doom for the English. By December, 2,260 had been lost to illness, and within a month Vernon and Wentworth, by now bitter enemies, withdrew from Cuba and were both recalled to England. The hero of Porto Bello's career in His Majesty's Royal Navy was over, and after a few more failed assaults by the English, so was the War of Jenkins' Ear.

KING GEORGE'S WAR

The War of Jenkins' Ear, fought for control of the Caribbean, overlapped with yet another war for possession of North America between the French and the British. As with most other French and Indian wars of the century, this one began with various raids on local villages, such as the May 23, 1744 assault by the French and Wabanaki Indians on Canso, Nova Scotia that netted over 100 prisoners of war. Yet as always, possession of Quebec and the maritime seaports became the focus of the war's major actions.

A joint British-American force laid siege to Louisbourg from May to June of 1745. After losing only 100 men, the English were able to force the surrender of the 1,800-man French garrison. Their celebration proved short-lived however, as pestilence proceeded to settle upon both the victor and vanquished. Within months 900 more British lay dead as well as 300 Frenchmen. Concerns about the ability to hold their prize quickly surfaced among the English and Americans.

A French relief force under the Duc d'Anville was launched in June of 1746. Composed of 11,000 men and 64 ships, the expedition was thought to be more than capable of retaking Louisbourg from the disease-weakened English. Yet storms began to slow the fleet, prolonging the length of the trip to three months. By September typhus and scurvy were reported as endemic aboard the French ships. Roughly 10% of d'Anville's army would die during the crossing of the Atlantic, with the Duc himself succumbing on September 27th. Conditions deteriorated so quickly that d'Anville's successor, d'Estourmel, attempted to commit suicide, and when that failed, resigned his post. After the fleet landed in the fall of 1746, 41% of the army followed those who had been lost in the crossing to the grave. By October 15th the returns of the army listed 587 more dead and 2,274 sick out of the 7,000 men that were left of the original large invasion force. Disease had preserved the English conquest of Louisbourg.

Worse was to come, however, as the various diseases visiting the French army soon began to spread inland. The Micmac Indians, allies of the French, began to die off from disease, soon after followed by an estimated two thirds of the Cape Sable Indian population. Finally, it has been suggested that 8% of the male population of Massachusetts succumbed as well, with the disease sparing neither white nor Indian.

The Treaty of Aix-la-Chapelle signed on October 18, 1748 accomplished little in the eyes of both sides. Louisbourg was returned to France and Madras, India to the British. Many colonists grew furious at this perceived betrayal of their efforts. In the end Americans had little to show from the wars of the 1740s save disease, death, and debt.

Overall, the various wars fought for the control of the Americas between the 1630s and 1740s altered little about the political landscape of the colonies. A century of violence had been accompanied by a century of pestilence and swept tens of thousands of men from the face of the earth. Though it was perhaps predictable that a final confrontation would take place between the English and French for dominance in the region, many within the colonies were already beginning to question the benefit to be gained from such a costly undertaking. Perhaps the thought of contemporary poet Thomas Gray in *Elegy Written in a Country Churchyard* best summed up the thinking of the time:

> *The boast of heraldry, the pomp of power,*
> *And all that beauty, all that wealth e'er gave,*
> *Awaits alike the inevitable hour.*
> *The paths of glory lead but to the grave.*

CHAPTER FIVE

"PESTILENCE GAVE THEM A COMMON DEATH"
DISEASE AND THE ENGLISH CONQUEST OF NORTH AMERICA

"Now, God be praised, I die contented"
—GEN. JAMES WOLFE, 1759

A s the previous colonial wars in North America had solved little, it was only a matter of time before a larger and final battle erupted for control of the continent. The catalyst for this renewed confrontation between the English and French centered on control of the fur-rich Ohio region. As previous wars had depopulated the French allied Indian tribes of the east and depleted the beaver reserves of the region, the various fur merchants had to move further west and inland in order to continue their trade. This movement collided with the penetration westward of English colonists seeking more land to cultivate. The quest for control of the Ohio region for divergent purposes would erupt in the 1750s in the final and most famous of the pre-Revolutionary wars.

THE FRENCH AND INDIAN WAR
The road to war began in earnest in 1753 when the French dispatched Paul Marin de la Malgue with 2,000 men to drive British trappers, merchants, and farmers from the region. Yet Marin's fort and road building expedition through the wilderness of western Pennsylvania took its toll upon the French. By October over 400 of the expedition's men including Marin himself had succumbed to disease, with most of the sick survivors being sent back to Montreal. The governor of New France, Michel-Ange Duquesne, upon seeing the returning soldiers, *"could not help being touched by the pitiable state*

to which fatigues and exposures had reduced them.[1] The initial French efforts to expel the British and quickly lay claim to the region had failed. Disease had ensured that the war that was to ensue would be a long and drawn-out process.

The first major British expedition of the war aimed at securing the Ohio region by force was likewise severely hampered by disease. English war plans called for an aggressive series of offensives in 1755 in order to seize major French-held posts including Fort Duquesne, Fort Niagara, Fort St. Frederic, and Fort Beausejour. The first of these was entrusted to Major General Edward

Young George Washington, by Charles Wilson Peale, 1772.

Braddock, who landed in Virginia in February at the head of two British regiments. Supported by colonial militia, Braddock advanced slowly through the wilderness towards the Monongahela River. Despite all its finery the force was ill-supplied, possessing mostly Indian corn, moldy biscuits, and salt pork for the troop's consumption, thus disease quickly erupted on the march. In this campaign a young Virginian officer named George Washington first showed the qualities that would serve him well in the harsh winters at Valley Forge. Stricken with dysentery and hemorrhoids, Washington was left by the roadside for two days to improve his condition and was forced to ride in a wagon during the uncomfortable march through the wilderness towards the French. In his letters and journals, the future president delicately referred to his, *"inveterate disorder in his bowels,"* as a *"pulmonary condition,"* and thanked, *"Doctr Jas Powder, wch is the most excelt mede in the W."* for his recovery.[2] At the battle that would erupt on July 9th and spell doom for the English expedition, Washington had to ride with a pillow between himself and his saddle in order to function effectively.[3] Though Braddock and the majority of his men would never return to Virginia, Washington's promising career had begun.[4]

Similar failures visited the other major British offensives of that year. Illnesses related to a lack of quality food and conditions began to plague the British and colonial armies. Brig. General John Forbes' expedition to capture Fort Duquesne once and for all, in 1758, was beset from early on by such diseases, including dysentery and scurvy. Its slow, methodical approach to the French fort only further worsened its already stretched supply lines, reducing the men to depredation. The French would eventually destroy the fort to prevent it from falling into British hands, depriving the English of ultimate victory. Forbes, who was himself sick with dysentery, would die shortly after his return to Philadelphia. British returns showed similar situations in other regiments, with 10 of them reporting 627 cases of scurvy alone in 1757. London ordered various units to farms in New Jersey in order to secure fresh provisions and improve troop health and morale. Where fresh fruit and vegetables were not provided, the local soldiers sometimes took matters into their own hands. Robberies were reported in the region of Albany in 1757 and Montreal in 1761 during its occupation, not of gold or valuables, but of local fruit. Conditions worsened to the point that in 1758 Gen. Amherst ordered the pillaging of local apple farms in Massachusetts to obtain provisions.

After Braddock's death, the overall British commander in North America became William Shirley, the Governor of Massachusetts Bay. The hero of Louisbourg a decade before, Shirley again planned a direct assault northward towards the heart of French Canada, first intending to take Fort Niagara. In order to facilitate this expedition, the Governor marched to and reinforced Ft. Oswego on the shores of Lake Ontario. Quick thinking by the French led them to assault and destroy the British supply depot at Fort Bull. Besides effectively crippling Shirley's invasion plans, the French assault also strained English supplies to Oswego's defenders. As the enemy began to besiege Ft. Oswego in August of 1756, the colonials inside were already battling scurvy and disease. Of the 2,000 men stationed at the fort, less than 1,200 were fit for duty to confront the French. Within four days Colonel Littlehales surrendered. One officer at the time expressed a gallows humor towards the benefits of perishing from disease over starvation. "*Had the poor fellows lived, they must have eaten one another.*"[5] The French under the Marquis de Montcalm had suffered less than 30 casualties and had pushed back the English lines from the shores of Ontario for two years.

Though initial British attempts to seize Canada may have failed, their movements did introduce another element into the battlefield: smallpox. For two years, from 1755 until 1757, the Great Canadian Smallpox Epidemic ravaged the French and their Indian allies, reducing their effectiveness in the war. By October of 1755 the Montagnais Indians living around Montreal were reported to be dying in large numbers from the pestilence. Within a few months the Seneca were also stricken. By 1757 Quebec was burying 20–30 people a day, with up to 3,000 hospitalized. By the end of the outbreak over 520 would die, more than had been lost to battle at that point and as many men as the British had lost at Braddock's disaster.[6]

The French were not the only ones to suffer from the outbreak of epidemics during the course of the war. By 1756 smallpox was in fact reported prevalent in a number of English regiments. The situation worsened to the point that the colonials began to fear the effects of introducing units and their diseases into local cities and towns. As a consequence, Philadelphia, when asked to provide housing for arriving British soldiers in 1756, refused. The situation was eventually resolved thanks to Benjamin Franklin, who proposed providing New Providence Hospital to the British to house their men, safely quarantining them from the general populace.

While Philadelphia may have been spared an outbreak of the plague, other regions weren't so lucky. Smallpox advanced up the Hudson River, moving town to town and fort to fort during the height of the war. Fort Edward, the vital English bastion commanding the access between the Hudson and Champlain Rivers, reported burying five or six men a day, with less than two thirds of its men fit for duty. By 1757 the disease had infiltrated Fort William Henry further to the north as well. This was unknown to the 8,000 French and Indians who emerged from the woods to surround the fort in August of that year. After a week of French artillery fire and ever approaching trenches, the British under Lt. Col. George Monro surrendered the post. A quick decisive stroke by Montcalm had handed him control of the upper Hudson and Champlain region. The massacre of the English prisoners committed by the French-allied Indians which followed is well known to history. Upwards to 200 Englishmen and colonials were scalped or simply killed by the various native groups as they withdrew under the terms of the surrender. Unfortunately for the Indians, Munro's revenge was much more effective than his resistance at William Henry. The scalps and clothing seized by the Native warriors and brought home as trophies of their great victory were contaminated with the same smallpox virus that had beset the fort. The disease proceeded to ravage the various groups, reducing their effectiveness in the upcoming campaign season. Likewise the prisoners deported by Montcalm to Canada by boat similarly spread illness among the French. In fact the various pestilences brought aboard by the English so enervated the crew that the prisoners themselves had to pilot the boats into Halifax to avoid wrecking. By the autumn of 1757 over 16% of the 18,000 English soldiers in North America were reported as sick with dysentery, typhus, or scurvy. In fact the continued spread of disease among the various French-allied tribes, especially upon their return home for the winter to their villages in the west, prompted many of them to remain home during the campaign seasons of 1758 and beyond. These manpower losses severely limited Montcalm's ability to prosecute the war.[7]

The English had attempted to capture the French fortress of Louisbourg on Cape Breton Island for a half century. As previously mentioned, the town had been taken with effort in 1745 before being returned to the French by treaty, much to the anger of the colonists who had given life and limb to take it. An additional attempt in 1757 during the current war failed due to a combination of weather and the presence of Admiral de la Motte off the coast. Yet once again disease reared its head, with typhus breaking out in the French fleet. Conditions quickly became terrible enough to require the French navy to return home. This allowed for a successful British assault on the fortress the next year and another expulsion of local Acadians.[8]

Negotiations with the Natives in Ohio, 1765 engraving.

Finally, as in previous wars, England again attempted to assault the sugar-rich islands of the Caribbean. One of the first campaigns involved an invasion of Cuba by the British in 1762. Thousands of Spanish soldiers succumbed to the Havana Yellow Fever Epidemic during the year leading up to the invasion, severely reducing the strength of the defenders. The massive English force landed in June of 1762, surrounded the city of Havana, and began a siege of the most valuable harbor in the region. The Spanish under Governor de Prado sought to delay the attackers, hoping that the various tropical diseases of the island would slowly reduce the English as had previously happened during other invasions. Though yellow fever did begin to ravage the British, the Earl of Albemarle was able to reduce Havana by August 13th. Once again, peace negotiations proved stronger than the guns of the enemy, as the entire island was returned to the Spanish in 1763. Yet in the meantime 15,000 British would succumb to disease during their occupation of the land, far more than had died securing it.

A similar occurrence took place during Commodore John Moore's expedition to Guadeloupe in 1759. Having landed on January 24th, the 6,000 British and American soldiers proceeded inland on a 5km march to engage the French. Within a week, 25% of the force was dead or dying of disease. Major General Peregrine Hopson, the commander of the land force, himself followed these soldiers to the grave on Feb. 27th. By the end of the month over 2,100 men had been evacuated due to illness, leaving less than half of the original force on Guadeloupe. Thanks to quick action on the part of Moore, His Majesty's Navy was able to reduce the French positions on the island and help the weakened British force to advance and secure one of the richest French possessions by May. Yet the occupation by the English of the island produced yet another 800 corpses before the end of the year.

Even the great triumph of the British during the war, the seizure of Quebec following the Battle of the Plains of Abraham, did not escape the tinge of pestilence. Wolfe's great victory came at the cost of only 60 Englishmen killed, not including his own great sacrifice. Yet the occupation of the city proved to be much deadlier. Disease and scurvy infected both soldiers and civilians in Quebec, resulting in the deaths of almost five times as many people as had died in the battle before its walls. The British were able to win this new battle by employing an old native trick, brewing spruce tea and spruce beer.[9]

In the end, despite losses by the British in both men and territory, the outcome of the conflict was never really in doubt. The military and demographic advantages of the English proved to be too much for the scattered and beleaguered population of New France. In addition, despite the name of the conflict, the role played by the various Native tribes on the side of the French was far less than had been anticipated by the English and feared by the colonists. Thanks to the almost constant barrage of illnesses that had been visited upon the Indians since the start of the Beaver Wars, the Native population of Canada was depleted and dispersed. Much as the diseases spread by Cortes during his retreat had paved the way for his victory at Tenochtitlan, so too did the previous successful and failed campaigns of the British eventually allow for their triumph in Quebec.

The French and Indian War and its associated disease-related depravity produced much more for America and England than just a list of casualties. England's expulsion of France from North America was truly a tremendous change in the geopolitical landscape of both Europe and America. Yet, the Treaty of Paris angered many on both sides of the Atlantic. England's return of Cuba to Spain and Guadeloupe to France, among numerous other exchanges, seemed to many in the colonies as a betrayal of the tens of thousands of men who had died mostly from disease in securing those lands.

A few positive changes did come out of the war in the handling of disease, however. Many English units stationed in North America began to grow food, not only to provide supplies but also to combat scurvy and other camp diseases as well. By the end of the war the 55th Regiment of Foot stationed at Oswego and the various units at Ft. Ontario reported bumper crops. Learning from the local Indians, other British units began producing spruce beer to combat scurvy, as has been seen in the aftermath of the fall of Quebec and Ticonderoga. All of these tactics would again be utilized by the English a generation later during the American Revolution, and represented the slow beginnings of the reversal of the impact that "General Pestilence" was having on the battlefield.

PONTIAC'S REBELLION

Though English domination was secured in North America following the Treaty of Paris, peace was not. The steady flow of colonists to the west of the Appalachians following the English victory began to be resented by the Natives inhabiting the region. The various tribes, the majority of whom had been allied with the French, had either watched with interest or actively participated in the French and Indian War. However, fears soon began to mount among them following the arrival of news that France was about to surrender the entire Pays d'en Haut region (inland, west of Montreal) to England.

From May to June of 1763 dozens of Native tribes in the Great Lakes area began to raid various English forts and American settlements in the region. Hundreds were massacred, tortured, or ritually cannibalized, while thousands more were pushed eastward towards the Appalachian Mountains. Those residents of western Pennsylvania who could not or would not evacuate to the east sought refuge instead at Fort Pitt. Built on top of the remains of the former Ft. Duquesne, the stockade's population had swelled to include over 500 civilians by the summer of 1763. Besieged for months, the commander of the fort, Capt. Simeon Ecuyer, began to worry not only about the hundreds of Natives outside his walls, but also the smallpox that was beginning to rage from within. A hospital was quickly established for the relief of the sick under the drawbridge of the fort, as dozens began falling ill.[10]

Growing desperate, Capt. Ecuyer eventually settled upon stratagem rather than strength to push back the Indian lines. On June 24th, Turtle's Heart and Mamaltoc, members of the Delaware tribe, approached Fort Pitt for a parlay. This was a tense moment for the garrison as Fort Miami had been taken by ruse when its commander Robert Holmes, who had been lured out by a native woman asking for medical help for one of her countrymen, was shot and killed.[11] At Fort Pitt, Indian agent Alexander McKee met with the two natives, and as was custom produced gifts for the men to take back to the larger confederation. Among these were two blankets and a handkerchief, all of which had been taken from smallpox victims in the English fort's hospital. Fellow Indian trader William Trent recorded the exchange in his diary: *"Out of our regard to them we gave them two Blankets and an Handkerchief out of the Small Pox Hospital. I hope it will have the desired effect."*[12]

Hundreds of miles away a relief column of 500 men under Col. Henry Bouquet was advancing from Philadelphia towards Fort Pitt. General Jeffrey Amherst wrote to Bouquet around the end of June asking, *"Could it not be contrived to send the small pox among the disaffected tribes of Indians? We must on this occasion use every stratagem in our power to reduce them."* The Colonel's reply penned on July 13th concurred, stating that, *"I will try to inoculate the bastards with some blankets that may fall into their hands, and take care not to get the disease myself."* The positive response from the commander only further encouraged the idea in Amherst's mind, *"You will do well to inoculate the Indians by means of blankets, as well as every other method that can serve to extirpate this execrable race."* All of these exchanges took place with no knowledge of Ecuyer's actions. By the time Bouquet had broken through and relieved the fort on August 20th, smallpox had already been unleashed against the natives.

The war would drag on for another three years. Cost and distance kept the English from achieving a full victory, while disease, defeats, and tribal desertions crippled Pontiac's progress. After thousands had been killed or captured the war ground to a halt with the signing of a treaty at Fort Ontario. Neither side gained from the war. Back in London, the costs of the French and Indian War and Pontiac's Rebellion prompted a radical solution to the issue of defending the colonies, the Royal Proclamation of 1763. Through this act, the British effectively barred unrestricted American settlement in Indian lands west of the Appalachians. Government officials hoped this would mitigate future conflicts between the colonists and the Natives, and thus reduce the cost for the British to maintain their empire and the fur trade. Needless to say, American colonists who had seen thousands of their countrymen dislocated, killed, or worse during a century of warfare with the French and Indians felt betrayed.

The smallpox unleashed among the natives has become a much de-bated topic. While it is well recorded that the pestilence did decimate the region's population during the war, it is impossible to say that its source was the biological attack launched by Fort Pitt. Undoubtedly the Natives did acquire the disease from local white settlers, much as had follow-ing the attack on Fort William Henry in the previous war.[13] Regardless of its source though, the attempt became both controversial and legendary, inspiring study and imitation for centuries afterwards.[14]

Nor was this the only contagion to be unleashed in the region by the English during the 1760s. Colonists moving up the Mississippi had brought malaria to the area during the years preceding and following Pontiac's Rebellion. The great explorer Alexander Henry, who traversed the region from 1760 to 1764, writes often of the mosquitoes he encountered, but not of malaria.

> Mosquitoes and a minute species of black fly abound on this river, the latter of which are still more trouble-some than the former. To obtain a respite from their vexations we were obliged at the carry-ing-places to make fires and stand in the smoke. . . . Mosquitoes and black flies were so numerous as to be a heavy counter-poise to

the pleasure of hunting. Pigeons were in great plenty; the stream supplied our drink; and sickness was unknown . . . The mosquitoes were here in such clouds as to prevent us from taking aim at the ducks, of which we might else have shot many.[15]

Yet by 1766, epidemics of the illness were reported at both Fort Chartres and Fort Kaskaskia.

From the Month of June to the 1st of October last the Ague & Fever has been remarkably prevalent—Insomuch that of the Garrison & Inhabitants of Fort Chartres & Kaskaskia few have escaped being more or less afflicted therewith & altho 'tis not in itself Mortal yet the frequency of it must be the Occasion of other Disorders that are so—Insomuch that not a single Person Male or Female born at the Illinois of Parents of fifty Years of Age & very few of Forty—Neither has there been any French Native of the Country known to have lived to an old Age.[16]

In 1768 alone, trader George Morgan records 50 men fit for duty and over 30 dead at Kaskaskia. The disease would continue and linger even after the rebellion. In 1796 the Frenchman Constantin-Francois Volney during his travels from Cincinnati to Detroit, a journey of over 700 miles, recorded less than 20 homes as being free from the illness. English diseases were transforming the Mississippi region in much the same way as the Spanish had done with Latin America.

Overall, disease helped to both conquer and secure North America for the English, determining the future of both the colonists and Natives. Pontiac's Rebellion in particular represented for the modern world an initial attempt at employing disease both actively and offensively rather than simply suffering through it passively in war. At the same time though, the British victory was achieved at a dreadful cost, both in treasure and lives. The colonists who saw themselves as suffering the lion's share of disease and death to secure acres for England were less and less convinced of the glory of the cause. As memories of the French and Indian War faded away, animosity began to fill the void. Disease was both to win and lose a continent for England.

CHAPTER SIX

TYPHUS AND TAXATION
DISEASE AND THE AMERICAN
REVOLUTION

"In this world nothing can be said to
be certain, except death and taxes"
—BEN FRANKLIN, 1789

If disease is such an immense agent of change, then surely if one studies
the greatest transformational event in American history, the Revolution,
its distinctive footprints would emerge. A brief study of the time period
reveals that pestilence was not only present during these events, but in fact
bore a massive influence upon the course of American independence. From
the road to separation from England to the campaigns, battles, and personalities of the war itself, illness shaped and drove the creation of the United
States of America. Some men attempted to use disease as a means to push
policy while others sought to employ it as a weapon. More often than not
though, it remained an uncontrollable force, chaotically winding its way
through revolutionary history.

ROAD TO THE REVOLUTION

The decade preceding the American Revolution was a period of rapid
change and development in terms of the nation's history, a period that has
been well studied for centuries in an attempt to highlight every possible
cause that led to the separation of the Thirteen Colonies from England.
Historians have opined for years as to whether the Revolution was a politically, economically, or socially caused phenomenon, yet little attention has
been paid to the role played by disease in the process. Upon examining the
historical record, one finds that certain interesting occurrences in the realm

of pestilence did take place, which in their own small part had inordinately large impacts upon America's road to independence.

The ending of the French and Indian War, far from securing peace and plenty for the colonies, instead plunged them and England into economic depression. Years of war led to a doubling of Great Britain's national debt. The results of this would be twofold, producing an economic recession in the colonies, as well as necessitating a more vigorous and expanded tax policy by England. The hardships produced by both of these would help propel the colonies towards separation. Yet, the situation was both worsened and in a minor way caused by a variety of diseases that erupted at the same time.

The latter part of the French and Indian War saw the eruption of a smallpox outbreak in New England. It is unknown whether this outbreak is connected with the eruption of smallpox at Fort Pitt in western Pennsylvania which occurred at the same time, or as contemporary accounts reported, that it began in the Boston home of either Deacon Paul Crowell from clothing imported from the West Indies, or Reuben Rider who received a bale of cotton transported from the South.[1] Boston had not seen an epidemic of the smallpox scourge since 1722, at which time half of the city's population had fallen ill, and one sixth of those had never recovered.

Conditions had improved in the city by the 1760s in terms of fighting pestilence. "Pest houses" were constructed to help quarantine the infected, and almost two dozen physicians roamed the city providing inoculations to the citizenry. Approximately 4,977 people in the city would be inoculated during the outbreak of the 1760s, of which only one percent would die. At the same time almost 700 would contract the disease naturally, and would experience a death rate of 124, or around 18%. The town selectmen, during their daily meetings kept detailed records of those infected and those treated. As well, they attempted to ration supplies of food and timber among other essentials, in case of a prolonged outbreak.[2] Though the physical effects of the disease were slowly being mitigated as America evolved, its economic impact was still deadly.

One of the individuals affected by this outbreak was local business owner Paul Revere. The economic depression that hit New England at the closing of the war had damaged his silversmith business like many others. With the arrival of smallpox, it was crippled further. One of Revere's children fell ill with the virus in 1764 and by law he was required to inform the town select-men of the outbreak within his house.[3] Rather than sending his child off to a "pest house," and a likely death due to a lack of care and unsanitary con-ditions, Revere agreed instead to have his home quarantined. Boston author-ities placed flags outside his house to advertise the pestilence within and posted guards around the clock to make sure that the Reveres did not leave. Though the family made it through the episode, the months that passed further enervated the family's silver business. No trade could be conducted during the quarantine and by 1765 Revere's business was floundering.[4] He began to practice various side trades and very quickly fell in with the Sons of Liberty as his economic condition and those of the citizens around him worsened from both the pestilence and the demands for more taxes.

Revere was not the only individual within Massachusetts to be finan-cially affected by smallpox. Industry as a whole was damaged due to the deaths, quarantines, and work stoppages. The more well-off members of Boston society chose to retire to the countryside during the outbreak in order to avoid the disease, but their absence made it almost impossible for the city's tax collectors to operate.[5] It was at this unfortunate point that the British Parliament began to pass a series of revenue-raising acts to help fund the massive debt that arose from the French and Indian War. The first of these was the Sugar Act, which passed in 1764. Coming as it did at the height of the smallpox outbreak in the region, the tax further damaged the already fragile economy of New England.

England experienced its own share of diseases during this period. In fact many of the Prime Ministers during the era suffered from both phys-ical and mental conditions that would drastically shape the course of Amer-ica's push towards independence. Charles Watson-Wentworth, the Marquess of Rockingham, who served as Prime Minister from 1765 to 1766, was a friend of constitutional rule in the colonies, going so far as to push for a repeal of the dreaded Stamp Act. Yet his anxiety disorder made him so inef-fectual in pushing his ideas through Parliament that he was replaced by the King after only a year in office. In fact Rockingham often resorted to drugs, alcohol, or bleeding in order to calm his nerves before important speeches.[6] Had he taken a more active role in mediating the conflicts both within Par-liament and between England and her colonies, the road towards America's independence may have emerged quite differently.

The notorious Stamp Act itself, which dragged America closer to rebellion, arose and fell due to an unlikely combination of medical issues. Another friend of American interests at the time was William Pitt, 1st Earl of Chatham who dominated Parliament from the 1750s until his death in 1778. Yet Pitt suffered from a series of physical and mental maladies that would hamper his effectiveness at certain critical junctures. One of these instances occurred during the debates for the passage of the Stamp Act in 1765. The Prime Minister had suffered from the gout since an early age. In fact when he was only 14 years old his physician, Dr. Arthur Addington, had prescribed a glass of port a day to Pitt to help mitigate the illness. The future prime minister readily followed the good doctor's advice, despite the fact that port is notorious for its toxicity and most likely simply magnified the problem. PM Henry Addington, son of Pitt's physician, once famously opined that, *"Mr. Pitt liked a glass of port very well, and a bottle better."* Completely crippled by an outbreak of the gout in 1765, Pitt was not present to argue down the Stamp Act motion, thus setting the stage for the slow slide of America and Great Britain towards war. With his recovery from the gout in early 1766, he was able to reappear before Parliament and worked to effectively repeal the Stamp Act that was causing so much consternation in the colonies.[7]

Yet within a few months, Pitt's health began to decline again. It has been suggested that the Prime Minister also suffered from a hereditary bipolar disorder, as he often reported hallucinations, an increased sensitivity to sounds, and mood swings.[8] Regardless of the cause, the flare up of 1766 effectively removed him from power in England during his administration from 1766 to 1768. Into this power vacuum stepped Charles Townshend, then serving as Chancellor of the Exchequer. Thanks to Townshend's aggressive efforts a new series of taxes, known in American history as the Townshend Acts, went into effect. These further enraged the colonists, whose refusal to pay and aggressive actions towards tax collectors prompted the Crown to dispatch thousands of soldiers in four regiments to Boston in late 1768. A little over a year later the famed Boston Massacre would occur, pushing the colonists ever closer to revolt. For his part, Townshend would not live to see the result of his Act, dying of typhus in 1768.

Rockingham and Pitt were far from the only members of the English government to suffer from debilitating mental and physical conditions which affected both their policies and the course of history. Perhaps no one figure from the period's health has been more carefully analyzed and misanalyzed than that of George III himself. The king suffered from periodic bouts of madness beginning in 1765, which would culminate in the famed Regency Crisis of 1788. In this episode, which followed a physical attack by George III upon his son, he collapsed into delirium, and Parliament negotiated removing him from power.

A similar disorder has also been suggested to have affected all four of his sons, including the future George IV, and his granddaughter, Princess Charlotte, who would die from complications related to childbirth, possibly from a pulmonary embolism. Likewise, numerous direct ancestors of his, including Queen Mary of Scotland and James I, professed similar symptoms. Thus, though official outbreaks of the condition are reported in both 1765 and 1788, one could hypothesize the permanent effect of the condition upon the king's decision making capacity for the years in between. The King was known to have experienced bouts of depression as well following the death of his two young sons, Octavius and Alfred, to smallpox in 1783.

The exact condition confronting King George III is debated. An initial prognosis in 1855 blamed acute mania, while in 1941 M. Guttmacher proposed depressive psychosis as the cause of the king's strange behavior. Yet both agreed upon the fact that it was a hereditary mental illness that was troubling the monarch. The 1960s saw a revolutionary diagnosis emerge to explain George's condition: porphyria, a group of diseases that, among other things stains one's feces and urine a royal purple. A variety of authors began

A Kick at the Broad-Bottoms, by James Gillray, 1807, showing the insanity of King George III.

to propose this rare blood disorder as the underlying cause of the king's erratic behavior.[9] Yet even this diagnosis left more questions than answers. The late onset of the condition in the king, combined with its severity when it manifested, led some researchers to look for a more proximate cause or catalyst. Research by Prof. Martin Warren has proposed arsenic poisoning as the actual cause of George's porphyria. Careful analysis of hair remains has revealed 300 times the toxic level of arsenic in the King's body. Though his wigs were powdered with arsenic, the massive amounts of the element in his blood were more likely the result of the King regularly ingesting James' Powder, a contemporary medicine that contained antimony.[10] In effect King George III was slowly poisoning himself over the course of his life. While this does not detract from a hereditary illness that almost certainly affected the monarch, his condition was most likely exacerbated by his own self-poisoning. The effects of this madness upon the decision making of the throne during the American Revolution is however, open to debate.

THE REVOLUTION: LEXINGTON AND CONCORD

As the First Continental Congress came to a close in 1775 it urged all colonies to begin raising and arming militias in anticipation of conflict with the Mother Country. As part of its planning, the Massachusetts Provincial Congress on February 21st, *"Voted, that Doctors Warren and Church be a committee to bring an inventory of what is necessary in the way of their profession, for the above army to take the field."*[11] Following inquires by these two gentlemen, on March 7th the Congress voted £500 for the procurement of various medical

Vaccination Against Smallpox or Mercenary and Merciless Spreaders of Death, by George Cruikshank, 1808.

supplies including 15, *"doctors' chests."*[12] These would be divided up the day before the Battles of Lexington and Concord to various locations around Massachusetts in preparation for war, including two at Concord itself.

The medical profession was heavily involved in the early days of the Revolution. The night before the famous "shot heard 'round the world," Dr. Joseph Warren, the then President of the Massachusetts Provincial Congress dispatched Paul Revere on his famous ride.[13] Accompanying the legendary patriot was another physician, Dr. Samuel Prescott. In fact, after Revere was taken prisoner by a British patrol, it was Prescott who made it to Concord to warn the townspeople of the approaching enemy force. Ten doctors would fight or serve in some capacity in the actions at Lexington and Concord the next day. Many Patriot lives were saved by their presence and the medical supplies provided by the Congress. Yet neither Warren nor Prescott could do anything to help John Parker, the famed leader of the minutemen at Lexington, whose famous order for the militia to *"stand your ground,"* came from a man confined to his bed the day before due to advanced tuberculosis. The disease would claim Parker's life only five months later.

SIEGE OF BOSTON

As the British soldiers fled from Concord and Lexington to Boston in April of 1775 they were followed by an ever-growing army of local militia. Yet, despite the size of the rebel force they could not possibly expel the entrenched English regulars from the city of Boston. For the next eleven months, the American units settled down to a protracted siege of the city, an operation shaped and impacted by the presence of disease.

Gen. Artemas Ward, who was kept from action at Ft. Ticonderoga in 1758 due to an attack of bladder stones, was initially in charge of the colonial units assembled outside of Boston from April to July of 1775. However, continuing bladder issues would keep him confined to a sickbed for a majority of the operation. And Ward was not the only one suffering from illness. Smallpox had broken out inside of the besieged city, and when combined with the ever decreasing amount of supplies available to the British, epidemic became a justified fear. On two occasions in late May, British detachments were dispatched to Grape Island and Chelsea Creek for fodder and livestock to combat their declining situation. Both times, Ward countered these moves with American assaults in an attempt to deprive the English of fresh provisions and worsen the conditions within the city.

Meanwhile, smallpox proved to be the largest danger though that was confronting the rebel army. As the majority of the British soldiers besieged in the city would have most likely experienced the disease back in England, they had slightly less to fear from it then the colonials. To help mitigate this danger, the Provincial Congress on June 27th,

> Ordered, that the committee to provide hospitals for the army, be directed to provide another hospital, to be appropriated solely for such of the army as may be taken with the smallpox, and to consider what measures may be taken to prevent the spreading of that distemper, and that Dr. Rand and Dr. Foster be added to the committee.

As the Revolution became a national issue, the Continental Congress promoted similar health provisions for the army.

> IN CONGRESS, Thursday, July 17, 1775. RESOLVED, That for the establishment of an Hospital for our ARMY, consisting of 20,000 men, the following officers and other attendants be appointed, with the following allowance and pay. A Director-General and Chief Physician, his pay 4 Dollars, per day.

Upon taking over command of the army on July 3rd from the ill General Ward, George Washington conducted an inspection of not only the rebel fortifications, but the recently commissioned hospitals as well. He wrote to Congress on July 21st:

> I have made inquiry into the establishment of the hospital and find it in a very unsettled condition. There is no principal director, or any subordinate among the surgeons; of consequence disputes and contentions have arisen, and must continue until it is reduced to some system. I could wish it were immediately taken into consideration, as the lives and health of both officers and men so much depend upon due regulation of the department. I have been par-

ticularly attentive to the least symptoms of smallpox, and hitherto we have been so fortunate as to have every person removed so soon as not only to prevent any communication, but any alarm or apprehension it might give in the camp. We shall continue the utmost vigilance against this most dangerous enemy.[14]

Washington soon began to appreciate the various diseases which were ravaging their foe inside of Boston. On July 27th he wrote to his brother that, *"the enemy are sickly, and scarce of fresh provisions."*[15] Seeking to build upon the actions of Ward, Washington proceeded to drive all remaining livestock inland in order to worsen the conditions the enemy faced. However the English were not the only ones suffering from illness. By the time Ward was relieved of command, his army was riddled with jaundice, diarrhea, various respiratory infections, pleuritic disorders, malaria, dysentery, and typhus. In total, around 15% of the army were reported to be too ill to fight.[16]

At the same time, American soldiers who had been captured by the enemy were being held in conditions that fostered both hunger and contagion. Gen. Washington wrote to his opposite number in the besieged force asking for better treatment for colonial prisoners, as they were, *"languishing with wounds and sickness."*[17] While General Gage did not appear to heed Washington's requests, he did seem to perhaps appreciate the danger that smallpox posed to the colonial militia.

Refugees began to stream in ever-larger numbers from the city as winter descended upon Massachusetts. Rumors began to circulate through the American rank and file that the English besieged in Boston would attempt to introduce the smallpox epidemic then ravaging Boston into the camps of the American rebels surrounding the city by releasing infected prostitutes or runaway slaves. Four British deserters who appeared in the rebel camp on December 3rd appear to have confirmed these accounts. Robert H. Harrison, aide-de-camp to Washington, reported the rumor to the Massachusetts Council in December . . .

SIR: I am commanded by his Excellency to inform you that four deserters have just arrived at head-quarters, giving an account that several persons are to be sent out of Boston, this evening or tomorrow, that have been lately inoculated with the small-pox, with design, probably, to spread the infection, in order to distress us as much as possible. The men are sent for their examination upon oath, who will give you such intelligence as may make it necessary to send down some judicious person to Pudding Point, where those people are to be landed, to examine into the matter, upon whose report proper measures may be taken to frustrate this unheard-of and diabolical scheme. Enclosed you have also a letter from Colonel Baldwin, containing the same account.[18]

Washington himself was skeptical at first of the idea, but soon reported to Congress that . . .

The information I received, that the enemy intended spreading the small-pox amongst us, I could not suppose them capable of. I now must give some credit to it, as it has made its appearance on several of those who last came out of Boston. Every necessary precaution has been taken to prevent its being communicated to this army, and the General Court will take care that it does not spread through the country.[19]

Washington had likewise written to Dr. Joseph Reed informing him that . . .

The small pox is in every part of Boston—the soldiers there who have never had it, are we are told under inoculation & consider'd as a security against any attempt of ours. A third ship of people is come out to point Shirley—If we escape the small pox in this camp, & the country round about, it will be miraculous— every precaution that can be, is taken to guard against this Evil both by the Gen court & myself.[20]

Wild rumors even began to be heard of General Gage's plans to fire arrows dipped in smallpox pustules into the American camp. Fear of the contagion spreading through his army led General Washington to propose the extreme measure of dipping his correspondences in vinegar to sterilize them.[21]

Dr. Joseph Reed, by Charles W. Peale.

At the same time that smallpox was stalking the streets of Boston, scurvy began to erupt among the English soldiers due to the ever-increasing lack of supplies within the besieged city. The British though were adept enough to adopt the recommendations of Dr. Lind from a generation prior. Watercress was grown on wet blankets in order to provide some basic form of vitamin C to the soldiers to combat the disease.

Due to a combination of worsening conditions within the city, a lack of viable options, and the arrival in the American camp of the "Noble Train," or artillery which had been captured at Fort Ticonderoga, the British soon resolved to withdraw from the city of Boston. The American success at Ticonderoga was due in part to the 46th Regiment that garrisoned it being composed mostly of invalids who suffered from a variety of diseases.[22] Immediately upon their evacuation, Washington ordered *a thousand men (who had had the small pox) under command of Gen. Putnam to take possession of the heights.*[23] The general undoubtedly still feared the presence and danger of contagion in the city, and for good reason. Drs. John Morgan and John Warren were tasked with recovering any medical supplies left behind by the British. Yet what the two men discovered was that all the medicines left behind had been contaminated by the British with arsenic. On orders from Gen. Washington, Morgan confiscated all of the drugs and shut down the largest pharmaceutical business in the city, which being owned by an alleged Tory could not be trusted. The presence of disease at Boston, and British attempts to harness it, would bear an effect not only upon the outcome of the American Revolution, but future wars as well.

MEDICAL DEPARTMENT OF THE ARMY

It was not long before both the various colonial assemblies and the Continental Congress were endeavoring to establish medical branches for their armed forces. As the epicenter of the rebellion, the Massachusetts Committee of Safety became the first to act, voting on April 29th . . .

that Major Bigelow be applied to, to furnish a man and horse to attend the Surgeons and convey medicines, agreeable to their directions; that Dr. Isaac Foote be directed and empowered to remove all sick and wounded, whose circumstances will permit, into the hospital, and to supply proper beds and bedding, clothing, victuals, furniture, etc., and that this be sufficient order for him to draw on the Congress for supplies.

This was to be the first hospital specifically designed to accommodate the sick and injured from the Revolution. The houses chosen to serve as hospitals all belonged to notorious loyalists, including Lieutenant Governor Oliver and Rev. Samuel Cook. Many of the injured from both Bunker Hill and the siege of Boston would be cared for at these homes.

On May 8th the Committee of Safety next set up a standing committee to examine the various surgeons in the rebel force. Prominent among this group was Dr. Benjamin Church, who had championed smallpox inoculation in the 1760s and would go on to become the first Surgeon General of the nation. Shortly afterwards, Dr. John Thomas was commissioned on May 13th to acquire medicine for the armed forces around Boston, while on May 14th Andrew Craigie was made commissary of medical stores, tasked specifically with requisitioning bedding for the sick and wounded.

On July 27th a Medical Department of the Army was formally established by Congress. Its bailiwick included both the numerous army hospitals as well as the various regimental surgeons. Dr. Benjamin Church was chosen as the first Director of the Medical Department, a position akin to the modern day Surgeon General. Having been educated at both Harvard and in London, Church was seen as the most qualified candidate. However, he proved to be both unpopular and unsuccessful in overseeing the various surgeons. A stream of complaints soon arrived on the desk of General Washington from Church's various junior surgeons. These issues proved to be moot however, as Church was put on trial in October of 1775 for treason. The first Surgeon General of the United States Army was found guilty of corresponding with the enemy, providing information to General Gage about American dispositions.[24]

Congress selected John Morgan to replace Church as Chief Physician and Director General, a position he would hold from October of 1775 to January of 1777. Morgan had served as a surgeon during the French and Indian War before studying at the University of Edinburgh. In 1765 Morgan and Dr. William Shippen had founded the Medical School of the College of Philadelphia. Once appointed, Morgan set about reforming many of the hospitals and practices then in use, continuing the work of Church. Yet, as the Surgeon General's powers included inspecting hospitals and overseeing surgeons, Morgan, much like his predecessor, ran afoul of the various surgeons in the Army, eventually choosing retirement in 1777.

The third Surgeon General of the American Army was Dr. William Shippen, the Medical School of Philadelphia's co-founder. He had previously headed Continental Army hospitals in both New Jersey and around New York City throughout 1776, seeing much of the fiercest fighting in those early years of the war. Shippen had actually been one of the main conspirators against Morgan during his period of leadership and was keen to replace him. Once made Surgeon General though, Shippen performed little better than those before him. By 1777 Dr. Benjamin Rush was publicly attacking Shippen for his response to an outbreak of hospital fever in Philadelphia, associating its severity with his inaction.[25] Hundreds of men would be buried each week, which in Rush's estimation was due largely to the Chief Physician's failure to provide adequate room and fresh air for those fallen ill. In a letter to Gen. Washington dated December 26, 1777, Dr. Rush laid out the causes of the army's current health crisis, notably 1) overcrowding, 2) lack of provisions, 3) lack of fresh clothes and sheets, 4)

lack of officers, and 5) the presence of an incompetent and all powerful director.[26] Yet Rush lamented most of all Shippen's political connections to Congress, which preserved and protected him for another three years. It would not be until 1781 that the now vindicated Dr. Morgan and Dr. Benjamin Rush orchestrated his forced resignation and court martial for financial irregularities.

The position of Surgeon General started out full of promise. Yet infighting, treason, and fraud marred it during its early years. It

Dr. Benjamin Rush, by Thomas Sully.

would not be until the elevation of John Cochran in 1781 that a competent and unblemished physician filled the position. That is not to say that many talented doctors and surgeons did not serve in the army during the war, with two of the most notable being Benjamin Rush and James Tilton.

Dr. Rush was trained in medicine at the University of Edinburgh before returning to America in 1769. Representing Pennsylvania, he signed the Declaration of Independence in 1776 and was appointed to be Surgeon General of the Middle Department. As mentioned above, Dr. Rush quickly came into conflict with Dr. Shippen over the latter's running of the Army Medical Department. Blaming the doctor for both embezzling funds as well as poorly running the various hospitals in his care, he personally lobbied for his removal. Rush went as far as to label the hospitals of the American army as, *"the sinks of human life in an army."*[27]

While employed by the Continental Army, Rush published a series of directions aimed at exploring *"the art of preserving the health of a soldier."* He concluded this could best be done by focusing on five things: Dress, Diet, Cleanliness, Encampments, and Exercise. In terms of the first, Dr. Rush argued against the use of a rifle shirt, a piece of wardrobe praised by Gen. Washington for its practicality in campaigns fought during various seasons. Rush, as a follower of miasmatic thinking, argued that, *"the shirt besides accumulating putrid miasmata, it conceals filth, and prevents a due regard being paid to cleanliness."*[28] As to the diet of the men, the doctor advocated against the usage of liquor as it weakened the health of the men. Finally, he argued against exposing the soldiers to cold or wet, arguing against campaigning in the winter as he thought it would produce more sickness in the army.

Dr. James Tilton who finished his medical education just before the outbreak of the Revolution, worked his way up from the Delaware militia to running medical hospitals at Princeton and Trenton. In fact, Tilton is perhaps best remembered for his work on the subject entitled, "Economical Observations on Military Hospitals; and the Prevention and Cure of Diseases Incident to an Army." He also attacked the medical centers established by Shippen, arguing that more soldiers were, *"lost by death and otherwise wasted, at general hospitals, than by all other contingencies that have hitherto affected the army, not excepting the weapons of the enemy."*[29] Tilton argued that to improve the health of the American soldier, nine things were needed: 1) Discipline, 2) Avoidance of excessive exposure to heat, 3) Supervised play and amusement, 4) Cleanliness, 5) Clothing, 6) Proper Diet including vegetables, 7) Hardihood, 8) Skincare, and 9) Training of the Mind. As will be discussed below, his greatest active contributions came during his time with General Washington in New Jersey from 1779–1780.

THE INVASION OF CANADA

While the British were still held up in Boston, the rebel leadership devised a daring expedition, the purpose of which was nothing less than the conquest of Canada. While all of the English provinces had been invited to attend the Continental Congress, the outlying colonies of Quebec, Nova Scotia, Bermuda, and others did not reply or rejected the idea of rebellion. Many influential Patriots felt that amongst these, Canada was both the most likely to join in the rebellion, due to its only recent acquisition by the British, and was also strategically necessary to protect the Hudson River Valley invasion route so often travelled in the previous century of wars between England and France. Thus in June of 1775 a scheme was adopted for the invasion of Canada.

The officer put in charge of the expedition was Major General Philip Schuyler. An overly cautious solider, Schuyler took time raising and equipping his army before proceeding on a slow march up the Hudson that would take his 1,200 men to Ile aux Noix, not far from Montreal. Unfortunately the general fell ill due to bilious fever and violent rheumatic pains, and was forced to turn over command to Brig. General Richard Montgomery. The new general inherited a force of sick, scared men who been camped in swampy, malarial terrain for weeks outside Fort St. Jean on the Richelieu River. The Americans assaulted the fort for two months in tough terrain and horrible conditions. By the end of October, with victory not any closer, over 900 soldiers had to be evacuated to Ticonderoga due to illness. It would take until November for the British garrison to finally submit. The road to Montreal was open, but at a high cost in both men and time.

Over two hundred miles away, a second force under Benedict Arnold was trekking through the wilderness of Maine up the Kennebec River to invade Quebec. By the end of September, only a few weeks into his march, dysentery had already struck Arnold's force. As reported by Dr. Isaac Senter, who accompanied the expedition, *"At this time several of our army were much troubled with the dysentery, diarrhea, &c."*[30] The situation would worsen as the invasion force continued its march through the month of October.

> Monday, 16th. We now found it necessary to erect a building for
> the reception of our sick, who had now increased to a very formi-
> dable number, A block house was erected and christened by the
> name of Arnold s Hospital, and no sooner finished than filled. . . .
> In this they left a young gentleman by name Irvin, 13 a native of
> Pennsylvania, brought up a physician in that city, and serving as an
> ensign in the company under Capt. Morgan. The case of this young
> gentleman was truly deplorable. In the first of our march from
> Cambridge, he was tormented with a disentery, for which he never
> paid any medical attention. When he came to wading in the water
> every day, then lodging on the ground at night, it kept him in a
> most violent rheumatism I ever saw, not able to help himself any
> more than a new-born infant, every joint in his extremities inflex-
> ible and swelled to an enormous size. Much in the same condition
> was Mr. Jackson of the same company, and Mr. Greene, my mate.
> The last of whom was left at Fort Western. All these three gentle-
> men were afflicted with the same disease during the beginning of
> our march, nor would arguments prevail on them to use any med-
> icine. Flattered as they were that nature would relieve them, yet
> they for once were mistaken.[31]

Only half of Arnold's force of 1,100 men would reach their objective
by mid-November. Montreal would fall to Montgomery on November
13th, allowing his disease-diminished force to link up with Arnold's own
depleted army outside of Quebec for an assault upon the old French bastion
of Canada. The assault, undertaken during a blinding snowstorm on
December 31, 1775, would prove a terrific failure, costing Montgomery
his life as well as 500 Americans killed or captured. Had the army moved
quicker and been less crippled by illness, the entire province might have
fallen to America.

The American force instead settled down to a siege of Quebec, hoping
to either reduce the city or await the anticipated reinforcements from the
colonies in order to assault the fortress again. Yet smallpox had already broken out in the rebel camp, soon to be followed by pleurisy and pneumonia.

As a soldier at the time reported, *"The smallpox is all around us, and there is great danger of its spreading in the army."*[32] The divergent backgrounds of the soldiers, each bringing their own microbes to the campaign, when combined with poor camp sanitation, lack of food, and general fatigue, produced an ideal environment for contagion. Conditions in the American camp worsened as the year progressed. Rumors began to circulate that General Guy Carleton, the British commander inside Quebec, had sent the infection into the American camp, in much the same way that it was feared Gage had at Boston. Thomas Jefferson himself believed these reports, writing to Francois Soules, *"I have been informed by officers who were on the spot, and whom I believe myself, that this disorder was sent into our army designedly by the commanding officer in Quebec."*[33]

By May, command of the disease-stricken force had passed to General John Thomas, the above-mentioned doctor who had been tasked by the Committee of Safety with acquiring medicines at the start of the rebellion. He quickly ascertained that only half of his 1,900 men were fit for duty, and a third of these were about to have their enlistments expire.[34] Dr. Isaac Senter with the force wrote:

> The smallpox still very rife in the army's new troops few of them who had had it. I was ordered by Gen. Thomas who commanded, to repair to Montreal and erect an hospital for their reception, as well by the natural way as inoculation. I accordingly made application to General Arnold, then commanding in the city, and obtained a fine capacious house belonging to the East India Company. It was convenient for nigh six hundred. I generally inoculated a regiment at a class, who had it so favourable as to be able to do garrison duty during the whole time.[35]

Despite the best efforts of 22-year-old Dr. Senter, the force was too far gone to be of any further use in the Canadian theater. A council of war held on May 5th quickly voted to withdraw the army to Trois-Riviere. Here they were attacked on June 8th by a force under Gen. Carleton. The Americans were forced to withdraw, though 200 men who were too sick to retreat were abandoned to the British. General Thomas eventually succumbed to smallpox as well: *"The disease was so malignant that he was entirely blind some days before his death. And what is remarkable, he had in the course of his professional life, been familiar with the disorder, and uncommonly skillful in its treatment, and yet had never taken it either by innoculation or otherwise."*[36] In fact Gen. Thomas had been firmly against the inoculation of his men due to his concerns over adding more capable soldiers to the ever-growing sick list.

Gen. Sullivan was commissioned to replace Gen. Thomas, finding that, *"smallpox, famine, and disorder had rendered them almost lifeless."*[37] He soon withdrew the remaining Americans from Canada to Crown Point. John Adams would report the dreadful condition of the army: *"Our Army at Crown Point is an object of wretchedness to fill a humane mind with horrour; disgraced, defeated, discontented, diseased, naked, undisciplined, eaten up with vermin; no clothes, beds, blankets, no medicines; no victuals, but salt pork and flour."*[38] George Washington would write to John Hancock in July expressing his fear that the smallpox then crippling the army could spread to other militia units in the area, weakening our northern defense.[39] A hospital was quickly erected at Albany under the direction of Dr. Samuel Stringer in order to both care for the ill as well as to inoculate the remaining soldiers.

The American expedition had failed. In the end, a good deal of the blame for this can be placed upon the various diseases that beset both the force and its commanders. As John Adams famously opined,

> The smallpox is ten times more terrible than Britons, Canadians and Indians, together. This was the cause of our precipitate retreat from Quebeck; this the cause of our disgraces at the Cedars, I don' t mean that this was all: there has been want approaching to famine, as well as pestilence. . . . The small-pox! the small-pox! What shall we do with it? I could almost wish that an inoculating hospital was opened in every town in New-England. It is some small consolation that the scoundrel savages have taken a large dose of it. They plundered the baggage and stripped off the clothes of our men who had the small-pox out full upon them at the Cedars.[40]

Disease kept Canada outside of the future American republic.

THE FRENCH-SPANISH ALLIANCE

Rampant malaria in the Southern theater threatened not only the British but the American defenders as well. In a letter dated December 24, 1776, Jose de Galvez, the Spanish Minister of the Indies, expressed to Luis de Unzaga, the Governor of Louisiana, Spain's willingness to supply the Americans with not only military supplies but quinine as well.

> You [Unzaga] will be receiving through the Havana and other means that may be possible, the weapons, munitions, clothes and QUININE which the English colonists [Americans] ask and the most sagacious and secretive means will be established by you in order that you may supply these secretly with the appearance of selling them to private merchants.[41]

Also known as Jesuit's Bark, quinine had been used for over a century to treat malaria in Europe. An adequate supply of the medicine would certainly provide the Continentals with an advantage during fighting in the Southern Department.

Further help from its allies included a massive invasion planned for 1779 that would have landed a Franco-Spanish army onto the very shores of England. Since the signing of the Treaty of Aranjuez between France and Spain in April of 1779, a variety of expeditions had been planned to weaken England, but the most audacious was an invasion of Great Britain herself.

Though the French fleet had departed in June, the Spanish fleet would delay rendezvousing with their allies for over a month. In the meantime, scurvy, typhus, and smallpox began to ravage the French crews. By the time the Spanish did arrive, a distinct shortage of supplies was further wreaking havoc on the French fleet's health.

> On August 16, their sick were "at least equal" to the number of sound men. Their line-of-battle ships had many of them had 50 to 60 percent of their crews [out of combat] and the dead were flung overboard in such numbers that it is recorded that "the inhabitants of Plymouth ate no fish for a month."[42]

The invasion was effectively destroyed. In the end, more than 8,000 sailors would perish at sea, with little to show for it.[43] Much as in 1588, England had been saved by pestilence.

THE BATTLE FOR NEW YORK

Though the colonists had temporarily expelled British forces from the Thirteen Colonies, victory in the war and independence as a nation were far from certainties. In fact, as an English invasion force under Howe was approaching the continent, so too were a variety of microbes.

Gen. Washington had moved his army to New York City shortly after the fall of Boston, correctly reasoning that this area would be the target of any future British attack.[44] By late spring the American forces had constructed a series of trenches and redoubts on Long Island (the area of today's Brooklyn), hoping to prevent an English landing. However, before the British arrived to confront them, the dreaded camp fever typhus had already begun to assault the American lines. Dr. James Tilton, who was with the army, recorded:

> The ignorance and irregularities of our men in the new scene of life subjected them to numberless diseases. The sick flow in a regular current to the hospitals; these are overcrowded so as to produce infection, and mortality ensues too affecting to be described ... The Flying Camp of 1776 melted like snow in a field; dropped like rotten sheep on their struggling route home where they communicated the camp infection to their friends and neighbors, of whom many died.[45]

Gastrointestinal diseases spread quickly through the soldiers, rendering almost 10% of them ineffective by June, 17% by July, 25% by August, and 32% by September.[46] Worse yet, with an estimated 20% of the childbearing women in New York City working as prostitutes, a massive outbreak of venereal disease raced through the American army.[47]

An attempt to construct a suitable hospital for the rebel forces, as first recommended by General Charles Lee, was stopped short of completion due to the more pressing need for barracks.[48] For the majority of the campaign both sides would rely upon private residences as well as King's College for hospital space. This in fact turned out to be a healthier alternative, as the dysentery and typhus that would erupt among the armies tended to spread in crowded hospital settings.

The prospect of smallpox following the army from Boston was a real fear. Unfortunately not only was inoculation forbidden by law in New York City, but the practice could possibly cripple the entire Continental Army. Instead Washington relied upon the use of Montresor's (Randall's) Island as a quarantine center for the infected. One Dr. Azor Betts was caught practicing inoculation and was summarily tried and sentenced to prison by the NY Committee of Safety, while Washington issued the following warning:

> Any officer in the Continental Army who shall suffer himself to be inoculated will be cashiered and turned out of the Army, and have his name published in the Newspapers throughout the continent as an enemy and traitor to his country.[49]

When the British did arrive in New York Harbor in June of 1776, they fared better against the muskets of the Americans than against the local diseases. Dr. Johann David Schoepf, the chief medical officer of the Ansbach Regiment, recorded in his diary the dreadful conditions that beset the Hessians that summer while encamped on Staten Island. Dysentery and scurvy were widely reported, with half of the regiment listed as sick or dying by 1778.[50] One soldier's diary describes the camp as a place where, "*the mosquitoes were dreadful,*" adding yet another vector for illness among the troops.[51] The factors warranted the release of a general order from the British headquarters on July 28th which read in part, "*The General is pained to observe inattention to the digging and filling of vaults for the Regts &, the General directs camp colourmen of the several regiments to dig vaults and fill up the old ones every three days; and that fresh dirt be thrown in every day to the vaults, and that all filth in and about the camp be buried daily.*" Though the ideas were quite advanced and sound, unfortunately for the British, enforcement of the order proved to be lax.

Conditions for the Americans proved to be little better than for the British, as reported by Dr. James Tilton of Delaware:

> In the year 1776, when the Army was encamped at King's Bridge in the State of New York, our raw and undisciplined condition at that time, subjected the soldiers to great irregularity. Besides a great loss and want of clothing, the camp became excessively filthy. All manner of excrementitious matter was scattered indiscriminately

throughout the camp, insomuch that you were offended by a disagreeable smell, almost everywhere without the lines. A putrid diarrhea was the consequence. The camp disease, as it was called, became proverbial. Many died, melting as it were, and running off at the bowels. Medicine answered little or no purpose. A billet in the country was only to be relied on. When the enemy moved to the East River, our army moved to White Plains and left their infectious camp and the attendant diseases behind them. It was remarkable, during this disorderly campaign, before our officers and men could be reduced to strict discipline and order, the army was always more healthy when in motion, than in fixed camps.

Nor were these situations lost upon General Washington, who in a letter to John Hancock in September of 1776 reported that upwards of a quarter of the Continental Army was sick. *"In almost every barn, stable, shed, and even under the fences and bushes, were the sick to be seen."*[52] The general proposed that these invalids should be moved to Orange Town, north of New York City to allow them to recover, prevent the spread of infection, and raise morale.[53] This last point apparently weighed heavily on Washington's mind, writing to John Adams that the militia were more susceptible to camp diseases than the regulars. This in turn led to increased grumbling, and then invariably, increased desertions.[54] By the beginning of fall, the Americans reported 7,610 men sick out of 22,000, while the British Army, which was somewhat healthier, could field an estimated 25,000 men. A notable casualty in the American army was Gen. Nathanael Greene, one of Washington's most able and distinguished soldiers, whose illness kept him from serving at the Battle of Long Island at which the Americans were defeated.

Washington gradually fell back across Manhattan as General. Howe slowly pushed north. A flanking maneuver by the British would eventually force the Americans to withdraw up the Hudson River and out of the city. Howe then returned to New York City, taking Fort Washington on November 16th and netting over 2,800 prisoners. Over the next few months, the disease-ridden conditions of these men's captivity would eventually kill 2,000 of them. The main American army would fare little better, with Washington's fear of disease in his exposed condition around New York City being one of the catalysts for his retreat across New Jersey.

As 1776 drew to a close Thomas Paine's words seemed to be quite apropos: it was truly a time that had tried men's souls. It can be estimated that in that year of the war alone, while 1,000 Patriots had fallen in battle, close to 10,000 had died of disease. John Adams recorded that during the following winter close to 2,000 soldiers died in Philadelphia of various contagions and were buried in Potter's Field. *"Disease had destroyed ten men for us where the sword of the enemy has killed one."*[55] As recounted by Dr. Tilton:

> The Potter's field of Philadelphia bears melancholy testimony of the fatal effects of cold weather on the military hospitals in the fall of 1776 and succeeding winter. Instead of single graves, the dead were buried in large square pits, in which the coffins were placed in ranges, cross and pile, until near full, and then covered over.

THE EARLY WAR IN THE SOUTH

Since the settlement of Jamestown, the South had been associated with disease and hardship. Not surprisingly the campaigns fought there during the American Revolution would experience the same conditions. As the siege of Boston progressed, Lord Dunmore of Virginia set about to raise a regiment of former slaves in order to secure British rule of Virginia. The Ethiopian Regiment, as it became known, quickly grew from a few hundred soldiers to thousands, as more and more slaves escaped from plantations and joined the British, many in hopes of securing their eventual freedom.

Yet as the slaves arrived, so too did smallpox. Dunmore soon wrote to the British government how he had *"too few to stay off Virginia having lost so many by sickness."*[56] The royal governor was forced to move his force from the ships where they had been crowded aboard to Gwynn's Island, leaving 300 corpses behind him. By June of 1776 Lord Dunmore, *"had erected hospitals . . . and that they are inoculating the blacks for the smallpox."*[57] Half of his force had been wiped out: *"had it not been for this horrid disease, I am satisfied I should have had 2000 blacks; with whom I should have had no doubt of penetrating into the heart of the colony."*[58] The disease crippled his army and by August forced the Governor to withdraw to occupied New York City. Yet to the colonists the entire inoculation operation bore a far more sinister aim. Rebel leaders and plantation owners feared that Dunmore's goal was to release the blacks back into Virginia, thus spreading the infection among the whites, killing thousands, or among their fellow slaves, thus wrecking the economy of the southern colonies.[59] Though widely discussed and published at the time, it is difficult to say whether this was the Governor's actual plan. What is clear is that 1) smallpox bought the southern colonies a few years of respite before Cornwallis' invasion; 2) the escape and drafting of 12,000 black soldiers led Congress in 1777 to repeal its prohibition against freemen from joining the Continental Army; and 3) both of these factors served to further harden the view of those in the South towards the institution of slavery.

Dunmore's smallpox scare was soon followed by a more serious and physical threat to the southern colonies. Well acquainted by this time with the dangers of the diseases prevalent to their region, many of the colonists saw in these pestilences a possible ally. Many in the Carolinas reasoned that the British would not dare launch an invasion of the area during the long, hot summer. Even after an English fleet under Sir Henry Clinton arrived off of South Carolina in the spring of 1776, Congressman Richard Hutson went so far as to predict that Charles Town would be safe until the late fall, considering it impossible for the British to attack during the malarial summer.[60] As the enemy fleet of nine ships and over 2,000 men lay moored off the coast, Sir Henry Clinton expressed similar misgivings about the operation. *"I had the mortification to see the sultry, unhealthy season approaching us with hasty strides, when all thoughts of military operations in the Carolinas must be given up."*[61] When the attack did take place, the initial English assault failed against the famed palmetto log fortifications of the rebels. Clinton had the option to regroup and attempt another offensive against the city, yet the growing threat of fever spreading among his troops led him to withdraw the entire invasion force back to New York. South Carolina was to be secure for another three years.

For their own part over the next two years, South Carolina and Georgia tried three times to launch invasions of British Florida. All of the assaults proved to be futile. The eruption of fevers among the various militia forces reduced their numbers and drove their commanders to withdraw, often without firing a shot.

PRISONERS OF WAR

The American Revolution bears a terrible distinction in the annals of American wars. Over the course of eight years, more colonial soldiers would die in captivity, largely from disease, than from bullets, bayonets, or conditions in the field. In fact official estimates place the death toll aboard British prison ships alone at twice that of the various battles of the war. When the history of the internment of American POWs by the English is studied, what emerges is a story of a premeditated massacre. Over 30,000 Americans were held captive during the war in New York City, of which an estimated 10–15,000 died, a number that equated to .5% of the population of the Thirteen Colonies. This represented a higher percentage in terms of population than the number of American lives lost in World War II.

After taking New York City, the British found themselves in possession of a large and growing number of rebel prisoners. Initially they used the various churches, including the Middle Dutch Church on Nassau Street, the Lutheran Church at Frankford, the Huguenot Church, and the Friend's Meeting House, City Hall, and even the classrooms of King's College (later Columbia University) to house their prisoners. Very quickly though, these various buildings were filled up, with one church reportedly holding around 700 men, and the city's New Gaol holding 20 men per cell, including the famous Ethan Allen. The English next turned to the various sugar warehouses of the city, most notably the Livingston Sugar House at Crown Street (now Liberty Street). At five stories, the building was one of the largest in New York City, yet in a short matter of time it was filled to overflowing as well. Ethan Allen himself recorded the horrors of the prison:

> I have gone into a church and seen sundry of the prisoners in the agonies of death in consequence of very hunger, and others speechless and near death, biting pieces of chip. . . . The filth of these churches was almost beyond description. I have seen in one of them seven dead at the same time.[62]

Disease soon erupted in the sugar houses as well. An outbreak of yellow fever claimed many lives and further convinced the English that larger facilities were needed to hold their prisoners.

To handle the vast number of prisoners, the British turned to a number of prison ships that were anchored in and around Wallabout Bay. Among the hulks off the coast were the HMS *Jersey, John, Scorpion, Strombolo,* and

Conditions aboard the
prison ship HMS *Jersey*.

Hunter, with the first becoming the most notorious. The *Jersey* was laid down in 1736 and ironically first saw action with Vice-Admiral Vernon at Cartagena where thousands were lost to disease. She had been de-masted in March of 1771 to be used as a hospital ship in New York Harbor, but as the Revolution erupted she would ultimately end more lives than she would save.

Housing over a thousand soldiers at any given time, the ship included men captured from campaigns as far afield as Halifax and St. Augustine. Disease ravaged the prisoners held below deck, with typhoid, dysentery, yellow fever, smallpox, and scurvy recorded by those who survived. Perhaps their own words speak best regarding the conditions aboard the prison ships:

> The heat so intense (the hot sun shining all day on deck) that they were all naked, which also served thme well to get rid of the vermin, but the sick were eaten up alive. Their sickly countenances and ghastly looks were truly horrible; some swearing and blaspheming; some crying, praying and wringing their hands, stalking about like ghosts and apparitions; others delirious and void of reason, raving and storming; some groaning and dying, all panting for breath; some dead and corrupting. The air was so foul at times, that a lamp could not be kept burning, by reason of which three boys were not missed until they had been dead three days. One person only is admitted on deck at a time, after sun-set, which nec-

essarily occasions much filth to run into the hold, and mingle with the bilge water.[63]

Robert Sheffield of Stonington, Connecticut, who penned the above account was only onboard the ship for six days. Yet in that time he managed to experience all of the horrors offered by the prison vessel, losing 30% of his captured crew.

The death toll aboard the *Jersey* could reach over 10 a day. British guards would approach the hatches in the morning demanding the prisoners "turn out their dead."[64] As they were often refused permission to come above deck at night to relieve themselves due to fear of escape, excrement would pile up for feet around the hatches by morning. When combined with the lack of good food, poor ventilation, lack of medical care, poor sanitation, and the various preexisting conditions brought aboard by the prisoners, disease spread rapidly. As one prisoner wrote:

At the time I was on board, there were about 1,100 prisoners, no berths to lie in, or benches to sit on; many were without clothes. Dysentery, fever, pleurisy, and despair prevailed. The scantiness and bad quality of provisions, the brutality of the guards, and the sick pining for comforts they could not obtain altogether furnished the cruelest scene of horror ever beheld.[65]

Viewing the colonists as rebels, the prisoners aboard the hulks were not often afforded the treatment that would have generally befallen them in a European war. Many infamous episodes and characters arose during these years of imprisonment, with one of the most notable being the Provost William Cunningham. Stories state that he once boasted of starving 2,000 rebels to death by selling their rations. According to a contemporary report in the *New Hampshire Gazette*, what little food there was consisted of an *"allowance to each man for three days [of] one pound of beef, three wormeaten biscuits, and a quart of salt water. The meat they are obliged to eat raw as they have not the smallest allowance of fuel."*[66] According to various reports Cunningham finally confessed to ordering the secret executions of hundreds shortly before his own hanging due to a variety of crimes he committed in London after the war.[67]

The pestilences that began in the prisons and aboard the moored hulks quickly spread to the inhabitants of New York City. Smallpox, dysentery, and typhoid erupted by the winter of 1777. Hundreds of more victims were thus added to the already growing death toll of the prisons. Those able to do so evacuated the city, the rich for their country homes, the poor for sanctuary, and many to seek inoculation. The prisoners themselves in their desperation even sought to self inoculate using pins or bits of glass. One such prisoner was Thomas Dring of Newport, RI:

> On looking about me, I soon found a man in the proper stage of the disease, and desired him to favor me with some of the matter for the purpose. . . . The only instrument which I could procure, for the purpose of inoculation, was a common pin. . . . The next morning I found that the wound had begun to fester; a sure symptom that the application had taken effect."[68]

Nor was the situation at the time unknown to the Continental Congress. General Washington himself wrote to General Howe in 1777:

> I am sorry that I am again under the necessity of remonstrating to you upon the Treatment which our prisoners continue to receive in New York. Those, who have lately been sent out, give the most shocking Accounts of their barbarous usage, which their Miserable, emaciated Countenances confirm. How very different was their Appearance from that of your Soldiers, who have lately been returned to you, after a Captivity of twelve Months; And, whether this difference in Appearance was owing to a difference of treatment, I leave it to you, or any impartial person to determine. I would beg that some certain Rule of Conduct towards Prisoners may be settled; if you are determined to make Captivity as distressing as possible, to those whose Lot it is to fall into it, let me know it, that we may be upon equal terms, for your Conduct must and shall mark mine.[69]

Congress would even appoint a commissary to help raise supplies for the captured Americans. Future President of Congress Elias Boudinot would serve in this role from 1777 to 1778, and worked earnestly to alleviate the suffering of the prisoners and lessen their casualties.

In all, it is estimated that around 30,000 prisoners were held in New York City over the course of the war, with about a third of them dying in captivity, mainly from disease. The true numbers of those who died may never be known as most bodies were unceremoniously disposed of in New York Harbor. Years after the war, when the first memorial to the prisoners was built, the locals quickly gathered up *"near twenty hogsheads full of bones were collected by the indefatigable industry of John Jackson esq, the committee of Tammany Society, and other citizens, to be interred in the vault."*[70] With a hogshead equating to 63 gallons, this represented a considerable amount of remains for a quick expedition so long after the war. The memory of the prisoners died hard in American history. Famous poems, the most notable of which was penned by Philip Freneau, memorialized the event for generations. Yet as the relationship between the United States and the United Kingdom softened with the dawning of the 20th century, facts blurred into tales, and tales became questioned. But as John Adams once opined, facts are a stubborn thing. The sacrifice of the men aboard the prison hulks should not be forgotten, for, as someone else once stated, *"what they wrote in ink in Philadelphia, they wrote in blood in Brooklyn."* The Prison Ship Martyrs Monument, which still soars high into the sky over Fort Greene in Brooklyn, continues to provide mute testimony to the atrocity.

WASHINGTON'S WAR AGAINST DISEASE

George Washington's concern with hygiene can be traced to his youth, as evidenced by his *Rules of Civility* written in his young teens. His experiences in the war against England only further led him to appreciate the advice of Rush, Tilton, and various other colonial doctors, especially as he saw the size of his rebel force diminish daily through defeat, desertion, and disease. The Battle of Trenton did much to remedy the first two issues, but disease had to be combated through different tactics than battle. And though the majority of occurrences of disease in the American Revolution had been natural in creation, as has been seen in the actions of Gage at Boston and Dunmore in Virginia, this was not always the case. The English were not adverse to using pestilence to their advantage in this conflict. In fact in 1777, British officer Robert Donkin, who was writing in occupied New

York, once again proposed using smallpox to weaken or exterminate American rebels. *"Dip arrows in matter of smallpox, and twang them at the American rebels, in order to inoculate them; This would sooner disband these stubborn, ignorant, enthusiastic savages, than any other compulsive measures. Such is their dread and fear of that disorder!"*[71]

Washington became an ardent believer in both inoculation and military sanitation out of tactical necessity. Writing to the Surgeon General, Dr. William Shippen, shortly after the Battle of Trenton:

> Dear Sir: Finding the small pox to be spreading much and fearing that no precaution can prevent it from running thro' the whole of our Army, I have determined that the Troops shall be inoculated. This Expedient may be attended with some inconveniences and some disadvantages, but yet I trust, in its consequences will have the most happy effects.
>
> Necessity not only authorizes but seems to require the measure, for should the disorder infect the Army, in the natural way, and rage with its usual Virulence, we should have more to dread from it, than the Sword of the Enemy. Under these Circumstances, I have directed Doctr. Bond [Dr. Nathaniel Bond], to prepare immediately for inoculating this Quarter, keeping the matter as secret as possible, and request, that you will without delay inoculate all the Continental Troops that are in Philadelphia and those that shall come in, as fast as they arrive. You will spare no pains to carry them thro' the disorder with the utmost expedition, and to have them cleansed from the infection when recovered, that they may proceed to Camp, with as little injury as possible, to the Country thro' which they pass. If the business is immediately begun and favoured with common success, I would fain hope they will soon be fit for duty, and that in a short space of time we shall have an Army not subject to this, the greatest of all calamities that can befall it, when taken in the natural way.[72]

When one considers the general view of most Americans and most colonial governments concerning the operation at the time, Washington's order to forcibly inoculate an entire army was revolutionary. As of 1776 the

Council of Safety in Baltimore still forbade the practice among soldiers, largely to prevent an epidemic from spreading among the general population. Boston had only lifted its own ban shortly before the signing of the Declaration, at which opportunity Abigail Adams had herself and her children vaccinated.[73] A few miles away in Marblehead a riot had broken out three years before after a small inoculation hospital was constructed following an outbreak. For his own part, General Washington had convinced Martha to undergo the operation in May of 1776 in order to protect her should she spend any time with the perennially unhealthy American army. Dr. Johann David Schoepf would go on to praise the growing popularity of the practice in America as compared to a continued prejudice against it on the continent: *"The almost universal practice of the innoc of smallpox, whereby countless mult. Of children are saved from death, to which the unconq prejudice of our fatherland still continues to offer sacrifice."*[74]

To further bolster the health of the soldiers, Gen. Washington and other commanders issued broadsides calling for a strict adherence to Mosaic sanitary codes. Some of the recommendations included the airing out of bedding, avoiding sleeping on cold and damp surfaces, the frequent washing or changing of garments (Deuteronomy 23:12), the isolation of sick soldiers (Numbers 5:1-4), and the digging of necessities far from the main camp (Deuteronomy 23:13-14). Washington himself looked towards Moses as the epitome of a successful general: *"a great army of the Children of Israel . . . that continued forty years in their different Camps, under the Guidance and Regulation of the wisest General that ever lived."*[75] These moves all shadowed recommendations made by Dr. Tilton:, *"Officers therefore, should be very solicitous to protect their men, as well as themselves, from the dreadful effects of filth and nastiness."*[76]

Yet despite these initial moves by Washington, the Revolution continued to falter, and disease continued to be a menace to the Continental Army. As the Americans wintered at Valley Forge after another disastrous season of campaigns, its men began to wage a far deadlier battle. Close to 3,000 Americans would perish from disease over the winter of 1777–78 at Washington's encampment. This time, the general was not alone in his movement for improved sanitation and general inoculation.

One of the heroes of the Valley Forge was Baron von Steuben, long renowned for his efforts to train the American army and transform it from a loose collection of militia into a modern, disciplined force. Yet an over-

looked and vital component of von Steuben's training involved a program
to improve the health of the encamped army, some components of which
would survive for a century. The Prussian general's famous work on military training included as many chapters and sections devoted to sanitation
and illness as rank and file firing. In Chapter 18, von Steuben recommended
various tips for sanitation . . .

> Whenever a regiment is to remain more than one night on the
> same ground, the soldiers must be obliged to cut a small trench
> round their tents, to carry off the rain; but great care must be
> taken they do not throw the dirt up against the tents.
>
> One officer of a company must every day visit the tents; see
> that they are kept clean; that every utensil belonging to them is in
> proper order; and that no bones or other filth be in or near them:
> and when the weather is fine, should order them to be struck about
> two hours at noon, and the straw and bedding well aired.
>
> The soldiers should not be permitted to eat in their tents, ex-
> cept in bad weather; and an officer of a company must often visit
> the messes; see that the provision is good and well cooked; that the
> men of one tent mess together; and that the provision is not sold
> or disposed of for liquor.
>
> The officer of the police is to make a general inspection into the
> cleanliness of the camp, not suffer fire to be made any where but
> in the kitchens, and cause all dirt to be immediately removed, and
> either burnt or buried. He is to be present at all distributions in the
> regiment, and to form and send off all detachments for necessaries.
>
> The quartermaster must be answerable that the parade and
> environs of the encampment of a regiment are kept clean; that the
> sinks are filled up, and new ones dug every four days, and oftener
> in warm weather; and if any horse or other animal dies near the
> regiment, he must cause it to be carried at least a half mile from
> camp, and buried.
>
> The place where the cattle are killed must be at least fifty paces
> in the rear of the wagons; and the entrails and other filth immedi-
> ately buried; for which the commissaries are to be answerable.
>
> The quartermaster general must take care that all dead animals,

and every other nuisance in the environs of the camp, be removed.

Chapter 24 deals specifically with the care of those already ill . . .

There is nothing which gains an officer the love of his soldiers more than his care of them under the distress of sickness; it is then he has the power of exerting his humanity in providing them every comfortable necessary, and making their situation as agreeable as possible.

Two or three tents should be set apart in every regiment for the reception of such sick as cannot be sent to the general hospital, or whose cases may not require it. And every company shall be constantly furnished with two sacks, to be filled occasionally with straw, and serve as beds for the sick. These sacks to be provided in the same manner as clothing for the troops, and finally issued by the regimental clothier to the captain of each company, who shall be answerable for the same.

When a soldier dies, or is dismissed the hospital, the straw he lay on is to be burnt, and the bedding well washed and aired before another is permitted to use it.

The sergeants and corporals shall every morning at roll-call give a return of the sick of their respective squads to the first sergeant, who must make out one for the company, and lose no time in delivering it to the surgeon, who will immediately visit them, and order such as he thinks proper to the regimental hospital; such whose cases require their being sent to the general hospital, he is to report immediately to the surgeon general, or principal surgeon attending the army.

Once every week (and oftener when required) the surgeon will deliver the commanding officer of the regiment a return of the sick of the regiment, with their disorders, distinguishing those in the regimental hospital from those out of it.

When a soldier is sent to the hospital, the non-commissioned officer of his squad shall deliver up his arms and accoutrements to the commanding officer of the company, that they may be deposited in the regimental arm-chest.

When a soldier has been sick, he must not be put on duty till he has recovered sufficient strength, of which the surgeon should be judge.

The surgeons are to remain with their regiments as well on a march as in camp, that in cCountryase of sudden accidents they may be at hand to apply the proper remedies.

Von Steuben went on to emphasize the dangers of disease to an army, relating that the commander, *"must never suffer a man who has any infectious disorder to remain in the company, but send him immediately to the hospital, or other place provided for the reception of such patients, to prevent the spreading of the infection."* In fact as seen from the general's orders concerning disease in Chapter 24 above, the person who bears the most responsibility for the health of the army remains the commander.

Thanks in large part to the ideas and efforts of Washington, von Steuben, and others, mortality rates among the soldiers of the north dropped from 1778 until the end of the war. Dr. Tilton himself went on to praise the work of von Steuben, writing, *"When the Baron Steuben was appointed Inspector General, besides the muster of clothing, he introduced a number of salutary regulations, which contributed more to the health and comfort of the troops, than the utmost efforts of all the medical staff."*[77]

Dr. Tilton lent his own expertise to establishing a modern military hospital at Morristown, New Jersey during the 1779–80 campaign seasons. Hoping to correct the horrendous and disease friendly conditions of the pre-existing camp hospitals, Tilton applied the latest proposals from eminent doctors such as Dr. John Jones to construct his infirmary. Jones was a pioneer of American surgery; having been educated in France, he treated many of the Founding Fathers and wrote the most widely used field surgery manual of the century.[78] Tilton's primary concern was to reduce the spread of infection, thus his hospital was laid out in a way that paid attention to the latest miasmic ideas. A lack of windows accept on the southern, sun-facing side of the building, a large central fire, and the position that the patients lay in were all designed to combat the spread of infections by air. Further attention was paid to reducing overcrowding and separating those suffering from fever from those who were simply wounded . . .

The importance of separating those ill of fevers, fluxes, etc., from the wounded and such as have only slight topical affections, will readily be perceived. Many a fine fellow have I seen brought into the hospital, for slight syphilitic affections and carried out dead of a hospital fever.[79]

His hospital stands out as one of the first specifically built to structurally limit the spread of disease.

Increased sanitation when combined with mandatory inoculation programs produced a healthier and more effective fighting force. Without this, winning the Revolution would have been a much more difficult, if not impossible proposition.

THE FINAL BRITISH OFFENSIVE IN THE NORTH

England launched its last major offensive in the northern colonies in 1777. Lord North had succumbed to illness over the winter, a condition which was to keep him confined at the royal residency of Bushy House well into the spring.[80] Thus the strategy for the upcoming season was planned in large part by Burgoyne and Howe in America, generals with competing outlooks towards victory.[81] The final product entailed a three-pronged invasion aiming to split the colonies along the Hudson River. The rather large complications of the operation, together with the even larger personalities involved, combined to produce failure in two prongs of the invasion, most notably and spectacularly at Saratoga. Thanks largely to the latter victory in New York, an alliance was formalized with the French and the war turned in favor of the Patriots.

The command of the English forces in the Middle Colonies passed to Sir Henry Clinton by 1778. The new commander-in-chief proceeded to pull the remaining British soldiers from Philadelphia and march overland back to Manhattan. Fortunately for him George Washington was unable to offer effective organized resistance due both to his lack of supplies and to the current inoculation effort underway in his army.[82] In the end Clinton was able to make it back to New York, yet he was able to accomplish little once there both due to the syphoning off of soldiers to engage the recently arrived French in the Caribbean, and due to the emergence of pestilence in New York City during the summer of 1778. An exceptionally hot summer descended upon the New Jersey–New York region and with it malaria and various other diseases that escaped from the prison ships of the harbor. Lt. Montresor in his journal estimated that by July over 850 men were ill in the General Hospital, and that by August seven out of every hundred men who entered the building would die.[83] Nor were the men on shore the only ones falling ill. On August 29th, six British warships dropped anchor off of Staten Island and let ashore 2,000 sick men. By May of the next year, Vice-Admiral Arbuthnot arrived with his fleet in New York Harbor to reinforce the English naval presence there. Yet the typhus he brought with him proved to be less than a welcome ally. The disease soon spread through the city, sending over 6,000 to the hospital. The English presence in New York would remain until the end of the war, but would serve solely as a holding force, being far too weak to venture out.

Thanks in large part to these American successes, when Lord North had recovered from his illness and issued a peace proposal in the winter of 1777–78, Washington was rightly able to view it with contempt.

THE WAR ENDS IN THE SOUTH

The final chapter of the American Revolution played out in the Southern colonies, and though disease had been a supporting player in the Northern Theater, it was to be the driving force in the South. For a variety of reasons the English felt that an advance northward from Georgia would be both easier than further offensives in the Northern colonies and help to bring about an end to the war. In preparation for this new thrust a British force under Gen. Augustine Prevost captured Savannah in December of 1778. The next spring the English began a slow and methodical march north to Charles Town (today's Charleston) to seize the largest rebel seaport in the South. Prevost made it up to the very gates of the city before the dual prospect of American reinforcement and the onset of the summer sickly season convinced him to withdraw back to Georgia. Though his rearguard of 800 men managed to hold off an American assault at Stono Ferry, almost half of it was unfit for duty due to disease. Both sides waited out the next few months, the Americans hoping for French naval reinforcements, and the English hoping for better health. General Prevost informed General Clinton that illness both among his men and in himself was to blame for the failure to take Charles Town.[84] By the late summer 28% of the roughly

4,000 men in Savannah were reported as sick. Nor were the Americans faring much better. General Benjamin Lincoln himself even requested a recall due to his own fears of Southern illness: *"I hope my friends will not suffer me to be kept here long."*[85]

Gen. Lincoln eventually chose to remain with the army, as the arrival of a French fleet under Comte d'Estaing boded well for the American chances to take Savannah and end the war in the South. Over 5,000 American and French soldiers sat down to besiege Savannah once the sickly season was over in September of 1779. On October 3rd, after two weeks of siege, d'Estaing began a concentrated bombardment of the city. Yet as the days accumulated so too did the sick and dying onboard the French fleet from outbreaks of scurvy and dysentery. Due in large part to this, d'Estaing instead ordered a foolish assault upon Gen. Prevost's main defenses. After only an hour, one of the bloodiest battles of the war was over. Almost 20% of the Franco-American force was killed, wounded, or captured, with the British suffering just 40 killed.[86] A week later, d'Estaing abandoned the siege. The war in the South would continue.

The next year an additional British force under Henry Clinton landed at Savannah. With an army of around 14,000 men the English were easily able to capture Charles Town and Gen. Lincoln's entire army by May 12th. The Americans had effectively now been pushed back to North Carolina. Clinton was confident enough in the outcome of the campaign to return to New York City and leave behind him General Cornwallis with instructions to clear out the remaining Southern colonies. Yet the Americans had left behind their own fifth column. The English garrisons at Charles Town and Savannah became increasingly ill as the year wore on. Malaria, yellow fever, typhoid, dysentery, and even dengue fever were all reported in British records. The occupying medical staff under Dr. John McNamara Hayes became not only overwhelmed but sick itself, with much of the hospital staff having to be returned to England to recuperate. Constant requests from the occupied cities to Cornwallis for reinforcements weakened his offensive capabilities. The general himself wrote to Clinton back in New York that his biggest enemy was the *"terrible climate, which except in Charleston, is so bad within an hundred miles of the coast, from the end of June to the middle of October, that troops could not be stationed among them during that period without a certainty of their being rendered useless for some time for military service, if not entirely lost.*"[87] Disease likewise affected the ability of the English to attract Loyalists to their cause, as both fear of camp diseases and the actual spreading of it among the recruits reduced their numbers and effectiveness.

Cornwallis had begun moving his army towards the area of Camden, South Carolina due to information that held the region to be healthier than the disease-filled country along the coast.[88] Yet this proved to not be the case, as the 71st Highland Regiment which had been posted east of Camden, had two thirds of its men succumb to fever. A hundred of the sickened regiment members, proceeding under orders from Major Archibald MacArthur down the Peedee River for medical help, were captured by American militiamen. Similar rebel assaults on various sickened outposts severely challenged Cornwallis' hold on the region by August. Further news arrived that a large Continental force under the hero of Saratoga, Gen. Horatio Gates, was proceeding south towards Charles Town. Yet as Cornwallis rushed to his army at Camden, he discovered a third of the men to be too ill to fight. Weighing his options, Cornwallis chose to fight for fear that his sickened force would be picked off as it staggered back to Charles Town. Fortunately for the English, Gates' army was not as large as they were led to believe, numbering only around 4,000 men. On top of this almost half of the men were stricken with dysentery and would prove to be of little value in the battle.[89]

Yet the victory Cornwallis would win at Camden did not lessen his problems. His army was weaker than ever and was now encumbered with thousands of equally sick Americans. He quickly decided to remove the prisoners to Charles Town as Camden was, "*so crowded, and so sickly, I was afraid that the close place in which we were obliged to confine them might produce some pestilential fever during the excessive hot weather.*"[90] The English commander next marched his men to Waxhaws hoping that the climate and topography of the region would improve the health of his army. Yet the diseases appeared to move as well: "*They say go 40 or 50 miles farther and you will be healthy. It was the same language before we left Camden. There is no trusting to such experiments.*"[91] The planned invasion of North Carolina was consistently delayed due to the various pestilences confronting the army. Cornwallis' entire cavalry arm in fact was stopped at Fisher Creek due to the severe illness of Lt. Col. Banastre Tarleton. When the army finally did move into North Carolina, it did so in the face of increasing Patriot resistance. The arrival of various groups of "over mountain men" eventually led the Americans to launch a successful attack on a group of Loyalists under Patrick Ferguson at King's Mountain. After only an hour of combat the American unit managed to kill or capture over 1,000 enemy soldiers. Though Cornwallis was lambasted for not supporting Ferguson with reinforcements, his decision was due entirely to the illness of Tarleton. "*My not sending relief to Col. Ferguson . . . was owing to Tarleton himself: he pleaded illness from a fever, and refused to make the attempt, although I used the most earnest entreaties.*"[92] Considered one of the most important battles of the war, King's Mountain effectively destroyed the Loyalist threat in the South and breathed new life into the Revolution. Cornwallis, himself ill for weeks after the battle, eventually ordered a general retreat back towards South Carolina.

The winter spent by the British camping at Winnsboro did much to improve the health of the army. Yet the damage caused by the various diseases had already been done. The various campaigns of 1780 can perhaps best be summed up by Josiah Smith, who wrote, "*Since the capitulation of Charles Town . . . a scene of devastation & distress has marked the route of both armies & their various detachments that I may in truth say, the sword, the pestilence and fire hath ravaged our Land.*"[93] With the onset of 1781 Cornwallis advanced through North Carolina, colliding with a rebuilt American army under Gen. Nathanael Greene at Guilford Court House on March 15th. Though the British army triumphed in the battle, it was a Pyrrhic victory. With his army reduced to just over 1,500 men in fighting condition Cornwallis had to plan his next move carefully. Gen. Greene decided to march into South Carolina to confront the English troops left behind under Lord Rawdon. While Clinton expected Cornwallis to go to the aid of his subordinate, Cornwallis instead proceeded northward into Virginia. The general would claim that strategy necessitated his reduction of this more powerful colony before the South could be secured. Yet perhaps his more urgent thinking was his "*hope to preserve the troops, from the fatal sickness, which so nearly ruined the army last autumn.*"[94] Cornwallis understood that without both more men and healthier men the war in the South would be lost, and perhaps Virginia's cooler climate would be a more ideal spot in which to rebuild his army. What followed was the brief campaign that culminated in the Battle of Yorktown, the surrender of Cornwallis and his entire army, and the practical end of the American Revolution.

When Cornwallis moved into Virginia he hoped for rest and reinforcements. The prior English commander in the region, General William Phillips, had died only a few days before from either typhus or malaria, a subtle portent of what was to come. Hessian soldiers sent to the Chesapeake to reinforce the British reported, *"Here dysentery and typhus fever were rife owing to the heat, and the scarcity and poor quality of the food."*[95] Yet Cornwallis continued to advance and even tried to use disease as an ally in this new campaign. After raiding the home of Thomas Jefferson, the governor of Virginia, Cornwallis took 30 slaves to infect with smallpox and use as biological weapons; 27 would die.[96] Jefferson himself wrote that the English General, by freeing them, *"would have done right, but it was done to consign them to the inevitable death from smallpox and putrid fever."*[97] The future president himself estimated that 30,000 slaves were taken from Virginia during the course of the war, of which perhaps 27,000 died of smallpox or fever.[98] Even once he was confined to Yorktown, Cornwallis continued his attempts to use pestilence against the American and French besieging him. In October, only weeks before surrendering, the English commander had sent hundreds of smallpox-ridden blacks from the town in an attempt to infect the Americans surrounding him, in a similar fashion to what Gage attempted at Boston. Benjamin Franklin recounts the event in his autobiography:

> Having the small-pox in their army while in that country, they inoculated some of the negroes they took as prisoners belonging to a number of plantations, and then let them escape, or sent them, covered with the pock, to mix with and spread the distemper among the others of their colour, as well as among the white country people; which occasioned a great mortality of both, and certainly did not contribute to the enabling debtors in making payment.[99]

As defeat seemed certain, General Cornwallis resorted to throwing the bodies of horses, steers, and slaves into the wells of the town in an attempt to cripple the American army once it occupied the area. Hearing rumors of these actions, Washington issued orders restricting interactions between his soldiers and the locals. Yet in the end, all of these nefarious actions did not save Cornwallis. On October 19, 1781, the entire English force at Yorktown surrendered and ushered in the ending of the American Revolution.

The English had been chased from the South not by the guns of the Patriots but by the proboscis of the Anopheles mosquito.

CONCLUSION OF THE WAR

Though the French and American soldiers who stormed the defenses at Yorktown had achieved a tremendous victory, the war needed longer to officially conclude. British units still occupied Charles Town, New York City, and many other forts and areas of the Colonies. The war would continue for another two years before the signing of a peace treaty at Paris in 1783. Much of this delay was caused not by tense negotiations but by the illnesses of a few key men.

Peace negotiations had commenced under Prime Minister Rockingham in 1782. Charles Watson-Wentworth, 2nd Marquess of Rockingham, had been elevated to the position of prime minister to replace Lord North whose government had fallen after the loss at Yorktown. Rockingham was a Whig who was seen as more sympathetic to the independence of America. He had previously been Prime Minister from 1765–66, during which time he had fought vigorously for a repeal of the Stamp Act. When he assumed office again at the end of March in 1782 he immediately contacted Benjamin Franklin and began negotiations for the total independence of America. Unfortunately less than 14 weeks later, Rockingham succumbed to influenza.

The negotiations did not die with the Prime Minister but they were crippled, as his replacement took a much harder line with the Americans. The new prime minister, Lord Shelburne, had been initially opposed to unconditional independence as a precondition for negotiations with America. The chief American diplomat Benjamin Franklin's own battle with gout during the summer further delayed the peace talks.[100] The negotiations would drag on amidst much intrigue and double-dealing until November of 1782 when an initial agreement was penned. This treaty was finalized and approved by Congress in September of 1783 and officially ratified the following January. The American Revolution was over.

Casualty estimates for the war vary. It is often reported that about 8,000 Americans were killed in combat, while another 25,000 were wounded. Yet an amazing 17,000 are reported to have died from disease, twice the number who were killed by bullets.[101] At the same time, it has been claimed that over 100,000 colonists died over the period of 1775–82 from the various smallpox epidemics that ravaged the region during the Revolution.[102] Though no monuments exist to these men and women, their losses were no less severe nor important to the fight for freedom. Disease both impacted and drove the American fight for independence; it would likewise shape the organization and growth of the nation in the coming years as well.

CHAPTER SEVEN

A NATION FORGED IN GOUT AND EXPANDED BY VENEREAL DISEASE
A MEDICAL LOOK AT THE EARLY REPUBLIC

"I die hard."
—GEORGE WASHINGTON, 1799

T he period directly following the victory of the Continentals at York-town was hardly an era of carefree celebration. From 1781 well into the presidency of George Washington, the fragile nation faced many obstacles, which James Madison referred to as the, *"mortal diseases of the Confederacy."*[1] Whether political, economic, or social, coming as they did in such a volatile era most of these problems necessitated a military response. Yet overhanging many of these was the ever-present danger of disease. Pestilence would become a tool by which the nation would be created and tested.

THE NEWBURGH CONSPIRACY
The American colonies had emerged from the Revolution independent; to some, too much so. The Articles of Confederation that was adopted during the war had loosely joined the Thirteen Colonies into the semblance of a nation. Yet the power of the federal government was limited, as each individual state jealously guarded its power against interference by both other states and Congress. The structure had managed to guide the colonists through the Revolution well enough, but various issues which arose after the Treaty of Paris began to show its limitations. Inflation, foreign and local debt, the failure of the British to fully evacuate the colonies, and issues over western land boundaries were used by some to expose the limitations of

the Confederation structure. Yet at the same time, many within the states championed the very limited government that was enshrined in the Articles. This included the majority of the nation's small farmers and even such notables as General Washington. Without a radical change in the political winds, it was unlikely that a stronger government would ever be adopted.

The defeat of the British at Yorktown in 1781 and the undertaking of peace talks did not resolve many important issues in the American Colonies. The most important of these to many involved the failure of Congress for years to pay the soldiers and officers who had secured victory at such high personal cost. Robert Morris, Superintendent of Finance for the new nation, had stopped paying the army in 1782 as he felt the end of the war had arrived. Though officers were promised a lifetime half pension, fears began to spread through the encamped armies that Congress lacked both the funds and political will to keep up the payment schedule. As payments fell off, many within the army even hinted at possible violence in order to receive their pay.

A proposal from Congress to pay the complaining officers off with a lump sum payment was met with shock and disbelief by Morris, who resigned in protest at the end of January in 1783. The situation only worsened as various officers called a general meeting in Newburgh, New York to discuss actions the army could take to pressure the recalcitrant Congress. It has been suggested by some historians that the army was bent on launching a coup to seize power, while others are unsure what exactly was being planned. Regardless of the extent to which they were willing to go, only one man and his medical condition could save the fledgling republic.

As the officers gathered at Newburgh on the Hudson and General Horatio Gates gaveled the meeting to order, an unwanted visitor entered the building. George Washington was not expected or welcomed to the gathering due to his republican views and loyalty to Congress. Yet it was doubtful that his presence or even the message from Congress that he carried would do much to convince the officers to alter their course of action. What did save the republic from falling to a military coup was Washington's presbyopia.

Like many other Americans, Washington's vision declined as he aged. In addition, like so many others who suffer from presbyopia, his vanity kept him from pulling out his reading glasses in the presence of others. But as the General delivered his opening remarks, he noticed that few of those gathered within the room were swayed by his pleas; a different tactic needed to be adopted. Washington reached into his pocket and produced a letter that he claimed was from a congressman who was sympathetic to the plight of the unpaid officers. As he opened and unfolded the document he paused, strained to read the letter, and proceeded to pull his reading glasses from his coat. Few if any present had ever seen their great commander wear his glasses. Washington noticed the puzzled faces around him and said, *"Gentlemen, you will permit me to put on my spectacles, for I have not only grown gray but almost blind in the service of my country."* Though the contents of the letter were unimportant and quickly forgotten, the effect of Washington's performance had effectively silenced the officers. Few could disagree that their commander-in-chief had not sacrificed as much or more than those present, and yet asked for nothing in return.

Washington's actions prevented a coup. His subsequent pleas to Congress eventually led to substantive relief for the unpaid officers. Yet, several small further mutinies continued to erupt across the nation, with perhaps the most successful being in Philadelphia where Congress was forced to abandon the town and flee to Princeton, New Jersey. All of these incidents, when combined with what was occurring in Massachusetts at the same time, lent further weight to the calls by many, especially Alexander Hamilton, for a stronger federal government—one which could both pay its bills and better control the military.

SHAY'S REBELLION

John Hancock served as the governor of Massachusetts for over a decade. He was a politician in the finest tradition of the *populares* of ancient Rome, doling out money to build personal power while representing the common man in his struggle against the landed interests. Hancock was overwhelming elected as the first Governor of Massachusetts in 1780 over his more aristocratic opponent James Bowdoin, who was disparaged for not attending the First Continental Congress. The latter had actually been suffering at the time from *"a long continued Slow Fever,"* most likely tuberculosis, but this was lost on the voters.[2] For five years Hancock guided the state through the end of the American Revolution, though he was unable to resolve the economic bubble that was threatening to burst within the region. Massachusetts possessed the largest state debt after the Revolution, valued at nearly $5,000,000 without taking into account the interest assumed by those who accepted paper from the commonwealth during the war. Hancock and the Massachusetts Assembly not only resisted raising taxes to pay off the amount, but did not service the interest for years either.[3] Citizens who served in the war demanded payment, debt holders wanted returns on their bonds, and farmers decried the already high taxes.

As the situation deteriorated, Gov. Hancock, though at the height of his popularity, suddenly announced his resignation and refusal to run for another term. Claiming to be stricken with a severe case of gout, Hancock retired to private life, paving the way for James Bowdoin and the propertied classes to assume power in the state.[4] A chain of events was then put into place in which the debt holders who now controlled the Assembly pushed through higher taxes to fund the state's obligations. Local farmers rebelled, as they were unable or unwilling to pay and as a result their property was confiscated. Shay's Rebellion stretched from August of 1786 to June of 1787, and though few were killed the event dramatically altered American history.

The revolt by the masses of Massachusetts that effectively paralyzed the state for months was employed by some in the political class to demand a stronger central government. One of the roadblocks to this movement had always been General George Washington, and due to this he became the target of a letter-writing campaign by prominent Federalists throughout the nation. Henry Knox's letter of Oct. 23, 1786, represents quite well the way that the rebellion was cast in the public light to frighten reluctant leaders towards the centralization of power,

> The people who are the insurgents have never paid any or but very little taxes. But they see the weakness of government: they feel at once their own poverty compared with the opulent, and their own force, and they are determined to make use of the latter in order to remedy the former . . . we shall have a formidable rebellion against reason, the principle of all government, and against the very name of liberty. This dreadful situation, for which our government have made no adequate provision, has alarmed every man of principle and property in New England.[5]

Various other leaders including Alexander Hamilton and James Madison pursued Washington and others until eventually a general convention was called for in Philadelphia to address the unworkable Articles of Confederation.

For his own part, John Hancock not only miraculously recovered from his condition but was back in the governor's chair the next year, heading Massachusetts from 1787 to 1793. His case of gout, whether it was real or simply political theater, had paved the road to insurrection that drew Washington and others out of retirement and into Philadelphia to draft what would become the Constitution of the United States.

NORTHWEST INDIAN WAR

Apart from leading the nation through the Revolution, the only other accomplishment of note of the Confederation government was the drafting of the Northwest Ordinance. As a perquisite for joining the new nation the various states had to give up land claims to territories in the region beyond Pennsylvania. Congress then set about dividing up the land along the lines of the Enlightenment thinking of the day. The territory was divided up into squares states, counties, towns, and pieces of property which were then sold off to land speculators in the east. Unfortunately the various Native tribes populating the region determined to resist the slow encroachment onto their lands and came together in 1785 to form a confederation under Blue Jacket, Little Turtle, and various other leaders.[6]

Raids by both sides commenced along the Ohio and Miami Rivers, leading to increased casualties. The scattered American forts throughout the region began to report extensive sick lists from malaria and other illnesses. *"The garrison is very sickly—forty nine men are this day sick with the intermitting fever, & what chagrins me most is, that we have not an ounce of Bark, & the sick obliged to live on salt provision for the want of hospital stores."*[7] President Washington responded in 1790 by dispatching an expedition under General Josiah Harmar deep into Miami country.

Harmar's plan called for a two-pronged advance against the Miami village of Kekionga, with the main army under him advancing down the Great Miami River while a holding force under Jean Francois Hamtramck would move from Vincennes up the Wabash to distract the Natives. Very quickly, however, the latter's force fell apart as his troops were stricken with malaria and an outbreak of measles.[8] Mutiny forced Hamtramck to return to his base and left Harmar unsupported in the field. The small American force that did make it to the Indian settlement was hardly the battle-hardened army that Harmar had been led to believe. A large Miami force under Little Turtle soon attacked him, and after several days of battles and ambushes the Americans had suffered almost half of their force killed or wounded. Gen. Harmar withdrew to present day Cincinnati having suffered the worst defeat to Natives in American history up to that point.

President Washington organized another, larger expedition for the following year under veteran commander General Arthur St. Clair. The latter, who had trained as a physician under Dr. William Hunter in England, had fought with distinction in the Revolution and had served as President of the Continental Congress during the acquisition of the Northwest Territory. Now, near 55 years of age, he was tasked with avenging the defeat of Harmar and pushing back the Native tribes which were overrunning the territory.

St. Clair's offensive almost immediately began to falter soon after it was launched. Due to a number of factors, his expeditionary force departed in mid fall rather than in summer as originally conceived by President Washington. In addition, illness soon reduced his manpower in half from its original strength of 2,000 men, and his movement was further slowed by the need to cut paths in the forest and build redoubts along the way. St. Clair himself suffered extensively from gout which kept him confined to a bed for much of his time in the field. On November 4, 1791 his army was surprised by a force of perhaps 1,000 Native warriors under Little Turtle. After hours of intense fighting, the American forces fled in panic leaving behind hundreds of wounded and camp followers who were most likely, subsequently slaughtered. Over 900 of the men engaged in battle were either killed or wounded making it the worst military defeat in American military history.

General St. Clair almost immediately blamed a lack of supplies and the presence of disease for his defeat. Writing to Washington only five days after the battle he stated that, *"Neither were my own exertions wanting, but worn down with illness and suffering under from a painful disease, unable either to mount or dismount a horse without assistance, they were not so great as they otherwise would, and perhaps ought to have been."*[9] A later Congressional investigation would largely agree with the commander, placing blame for the massacre on the Department of War and miasma.

The war would drag on until 1794, at which point General "Mad" Anthony Wayne and his newly raised and trained Legion of the United States decisively defeated Blue Jacket at the Battle of Fallen Timbers. Though the battle itself was merely a skirmish, the presence of a reinforced American army in the region was magnified by the outbreak of malaria among the English soldiers in the various forts along the border. Colonel R.G. England at Fort Miami, who had only recently assumed command after Major Campbell was forced to retire due to illness, had written to his superior, *"I am much concerned to mention to you that the Detachment of Royal Artillery, & 24th Reg't are much reduced by the Fever they are very unequal to Garrison Fort Miami if any thing hostile should be intended."*[10] Similar reports from other forts in the area when combined with Wayne's offensive posture finally served to convince the English to sign Jay's Treaty and withhold further support to their Indian allies. Interestingly though, Wayne's show of force was more bluster than truth. His army had returned to Fort Defiance in Ohio following the battle and was almost immediately devastated by malaria. *"The soldiery gets sick very fast with the fever and ague, and have it severely."*[11] By the fall of 1795 almost 70% of his army was too ill to fight, and Wayne was forced to await the arrival of quinine and other medical supplies. The much feared Legion of the United States was hardly a threat to the English or Natives in the region. America's first war as an independent nation had come to a close, won as much by disease as by force of arms.

LOUISIANA PURCHASE

Sometimes the impact of disease on the military ventures of other nations effected the United States as well. Nowhere is this more evident than with the French expedition of General Leclerc to Haiti. The administrations of Washington and Adams had established a prosperous and secure nation, yet one that was quickly being divided by two nascent political parties. The Federalists favored a program of increasing industry, which necessitated the growth of cities, while Republicans preferred the expansion of the nation across the continent. Yet one area that concerned both groups was the Mississippi River. Not surprisingly therefore, Washington worked towards opening up the Mississippi, a move he accomplished with the signing of

Pinckney's Treaty in 1795. Securing friendly relations with Spain and trade rights to New Orleans, the document kept the peace for only a few years; for with the rise of Napoleon and the defeat of Spain in Europe, the situation in America quickly changed.

The Third Treaty of San Ildefonso, among other provisions, ceded the territory of Louisiana back to France. Napoleon was now in possession of over 800,000 square miles of land, an empire it had not seen in the New World since 1765. Unlike the French of a century before, Napoleon had grandiose plans for his new territory. Rumors circulated of his ambitions to recruit an army of former slaves and use them to capture the valuable British Caribbean islands. Others alleged that he intended to march on Canada or perhaps even invade the United States itself. Napoleon's dispatch of troops to New Orleans in 1801 only further heightened fears in the region. Yet before any of this could be attempted, it was necessary for the French to first secure the island of Saint-Domingue.

A slave revolt had erupted in modern-day Haiti in 1791 at the height of the French Revolution. Proclaiming liberty for all slaves, the rebellion proved to be a bloody affair, with over a hundred thousand dying over the course of the next decade. A British invasion of the island launched in 1794 proved to be a disaster. Yellow fever and other diseases crippled the forces of General Thomas Maitland, killing perhaps 25,000. Medical services did little to ease the plight of the army, with hospitals being little better than mortuaries. Soldiers often chose not to report illnesses to avoid being sent to the buildings from which so few returned. By 1798 peace was made to allow the remaining English to withdraw.

Once securely in power in Paris, Napoleon dispatched an army of occupation in 1801 of over 20,000 men to resecure the island and possibly return slavery to the region. Regardless of his later goals and actions, President Jefferson was initially in favor of the expedition as fears of the successful slave revolt spreading to America were rampant among Southern Republicans.[12] The French under General Leclerc, Napoleon's brother-in-law, were quite successful within the first few months of their landing. By May the leader of the revolting slaves, Toussaint L'Ouverture, even arrived in the French camp to talk peace.

However, the arrest, imprisonment, and death of the revolutionary hero, when combined with dissatisfaction with the French treaty, led thousands of Haitian rebels to take to the wilderness and continue fighting. Months of slaughter and skirmishes would follow, but it was another native ally that would decimate the French invasion force and establish freedom on the island: yellow fever.

Within months the disease had begun to savage the French army. In fact two Swiss units had mutinied in France rather than face the illnesses of the Caribbean, fears that soon proved to be true.[13] Leclerc in a letter to Napoleon described the situation in the first few weeks as, *"if my position has turned from good to critical, it is not just because of the yellow fever, but, as well, the premature reestablishment of slavery in Guadeloupe and the newspapers and letters from France that speak of nothing but slavery."*[14] The disease very quickly restricted the ability of Leclerc to wage war against the Haitians. Each day 200 to 250 men were entering the hospital, of which only 50 would ever come out.[15] One regiment alone of 1,395 men saw its effective strength reduced to only 83 soldiers.[16] By the fall around 20,000 of the 35,000 soldiers in Leclerc's army had died from the disease, and another 8,000 were confined to the hospital. Fifteen thousand of these had died in two months alone.

Happy were the French soldiers who died quickly; others suffered from cramps, aching heads that seemed about to blow up, and insatiable thirst. They would vomit blood, as well as a substance dubbed "black soup," then their faces turned yellow, and bodies were encased in malodorous phlegm, before death happily intervened.[17]

The French took to burying the dead at night to avoid news of the calamity spreading, yet soon enough this could no longer be avoided. On October 22nd General Leclerc himself was diagnosed with the disease. Though his doctors at first refused to identify yellow fever as the culprit, his symptoms argued otherwise. He even took the step of removing himself to Tortuga to escape the ailment, only to die on November 1st shortly after his arrival there. Leclerc was only 30 years old at the time, and left behind a young bride—Napoleon's younger sister, Pauline—and a few thousand men who would struggle on in vain under Rochambeau until the next year.

By late 1803 it was obvious to the French that the war had been lost, yellow fever having crippled their hopes of retaining Saint-Domingue. As well, Napoleon's hopes for a New World empire were shattered. Broke and disillusioned, the French emperor informed Robert Livingston, the American diplomat then in Paris, that he was seeking to sell Louisiana to the United States for $15 million. Livingston had arrived in France the year before with orders from President Jefferson to purchase New Orleans for $10 million. Though not authorized to agree to the larger deal, Livingston eventually agreed. For only three cents an acre America had grown enormously, gained untold riches, and expanded to the Rocky Mountains. Yellow fever had doubled the size of the nation.

LEWIS AND CLARK

Such an immense expansion of territory necessitated an equally legendary exploration of it. To accomplish this, President Jefferson organized the Corps of Discovery in 1803. Better known as the Lewis and Clark Expedition, after the two men assigned to lead the journey west, the force was equipped with weapons, food, trade items, horses, and the best medical training and equipment available at the time. The information gained from the exploration of Louisiana would prove to be invaluable during the Mexican War, and the subsequent populating of the region would help lead to numerous military conflicts over the next century.

Three hundred years of exploration, tales of what lay out west, and Jefferson's own personal dealings with illness, convinced both the President and the Corps' leaders that preserving the health of the expedition was paramount.[18] To begin, Meriwether Lewis was sent to the famous Dr. Benjamin Rush for basic medical training. Rush would train Lewis in some of the most advanced techniques of the time, including the use of mercury and bloodletting. In addition, he helped to prepare the medical supplies for the expedition including the following:

Turkish Opium (1/2 pound)
Laudanum
Medicinal Wine (30 gallons)
Dr. Rush's Bilious Pills (50 dozen)
Clysters (Enemas)
Syringes
Pewter Penis Syringes
Peruvian bark (quinine: 3,500 doses)

Jalap (purgative)
Glauber salts (sodium sulfide)
Niter (potassium nitrate/saltpeter)
Tartar emetic (1,100 doses)
Calomel loaded with mercury
Antimony-potassium
Tragacanth

The total cost of Dr. Rush's medical supplies was $90.69, almost $2,000 in today's money and four percent of the total cost of the expedition. The kit is notable for its various purging elements, notably the tartar emetics, jalap (short for jalapeno), clysters, antimony-potassium, and Rush's own bilious pills, which contained 50% mercury as well as chlorine and jalapeno. In addition the expedition was apparently well supplied to combat both malaria (with the Peruvian bark) and sexually transmitted diseases (with the penis syringes and mercury). These carefully made choices clearly indicate what Lewis, Clark, Rush, and Jefferson thought that the men would encounter while proceeding westward.

In fact, aside from a few other minor illnesses, the above were the dominant conditions that affected the members of the expedition. Dysentery quickly hit the men soon after their departure in May of 1804. *"The party is much afflicted by boils and several have deassentary which I contribute to the water (which is muddy)*," wrote Clark in June of 1804. The members of the expedition used Dr. Rush's pills, which they nicknamed *Thunder Clappers*, to help combat the illness. Though the effectiveness of imbibing mercury is doubtful, the mercury-laden feces that the explorers left behind them allow modern researchers to very carefully track the exact route the expedition took.[19]

Various other minor medical incidents occurred during the journey. In one case, the Native woman Sacagawea's young child Jean Baptiste, called Pompey by the explorers, was diagnosed by Lewis with an infection from teething. He soon recovered following the use of an enema and the application of onions. Lewis himself was shot in the buttocks, but suffered no serious injury. One of the Field brothers was bitten by a snake, but was quickly treated with a poultice of bark and gunpowder. Finally, though ticks and mosquitos were reported as troublesome by Clark in his journal, though they seem to not have produced any associated illness.

By far the biggest threat to the expedition stemmed from venereal disease. Not only was it expected that various members of the Corps would engage in sexual relations with some of the Native women they came across, but in some cases it was encouraged. Various tribes across the region saw sex as a means by which to settle differences between tribes or a way in which to obtain power. This is especially true of the Arikara and Mandan peoples of the Great Plains. Husbands would offer their wives, and fathers their daughters to Lewis and Clark's men, seeing sex with them as a conduit by which they could acquire the white man's power. Even York, a slave who accompanied the trek, was propositioned in this way by the Natives. According to Clark, he was offered four women in one night due to his perceived spiritual and physical power.

The natural result of these encounters was a rash of venereal diseases that plagued the journey. In March of 1805 the explorers recorded that the men were, *"Generally helthy except Venerials Complaints which is very Common amongst the natives . . . and the men Catch it from them."* Overall eight men would catch gonorrhea or syphilis and were treated with mercury and penile syringes to combat the diseases.[20]

> Goodrich has recovered from the Louis veneri which he contracted from an amorous contact with a Chinnook damsel. I cured him as I did Gibson last winter by the uce of murcury. I cannot learn that the Indians have any simples which are sovereign specifics in the cure of this disease; and indeed I doubt very much wheter any of them have any means of effecting a perfect cure. When once this disorder is contracted by them it continues with them during life; but always ends in decripitude, death, or premature old age; tho' from the uce of certain simples together with their diet, they support this disorder with but little inconvenience for many years, and even enjoy a tolerable share of health . . . the Chippeways use a decoction of the root of the Lobelia, and that of a species of sumac common to the Atlantic states and to this country near and on the Western side of the Rocky Mountains . . . these decoctions are drank freely and without limitation. The same decoctions are used in cases of the gonnaerea and are effecatious and sovereign."[21]

Perhaps the most famous of the expedition's patients was Sacagawea. *"Sah-cah-gah we a, our Indian woman is very sick this evening: Capt. C. blead her."*[22] She was reported to be suffering from pelvic pain, fever, anorexia, and delirium. Her condition appears to have only worsened as the Corps moved upstream. Ingestion of salts, opium, and the application of Peruvian bark to her abdomen seems to not have helped her condition. It was not until the 16th of June, a week after the onset of her condition, that Sacagawea's fever broke and her condition improved after she was given water from the Sulphur Springs near Great Falls, Montana. Clark continued the opium treatments and bark applications as well until his patient was fully recovered. Modern doctors have speculated that Sacagawea's condition was most likely Pelvic Inflammatory Disease, or PID. An illness caused by a sexually transmitted disease, possibly gonorrhea or syphilis. Sudden recovery from PID is not unknown; in addition it can result in temporary infertility which would perhaps explain Sacagawea not experiencing another pregnancy for several years.[23] Whether the famed Native woman acquired the disease from the dalliances of her infamous husband or from one of the other explorers is unknown.[24]

Yet despite dysentery, sexually transmitted diseases, snake bites, and the occasional bear attack, the expedition returned relatively unscathed. Only one casualty is reported to have occurred, with Charles Floyd dying of a burst appendix on August 20, 1804. The medical arts of Lewis and Clark undoubtedly saved numerous men on the journey west. These skills were also used to establish relations and broach trade with the Natives. Lewis personally treated a number of conditions among the Nez Perce Indians in exchange for 60 much needed horses. While Lewis and Clark showed that movement across the continent was possible, they also demonstrated that proper preparations could reduce the impact of illness.

THE WAR OF 1812

The War of 1812 is often termed America's Second War of Independence, reasserting as it did the nation's freedom. Yet would this second revolution copy the same disease-related disasters of the first? A generation had passed since the colonies broke away from England. Though military tactics and technology had advanced little in America, medical knowledge had grown by leaps and bounds. Many of these innovations would make their appearance during the War of 1812, but would they be enough to avert disaster?

The War of 1812 is traditionally taught as having arisen from two main causes: British support for Indians in the Northwest (today's Midwest), and the impressment of American sailors. These were slow-moving causes that set in place a chain of events that was to drag the two nations eventually into war. Though each cause involved a complicated series of events, disease once again lay at the heart of both causes.

The alliance between the British and the Native Americans of the Old Northwest had been in place since the time of the American Revolution. The emergence of the United States as an expansionist nation stretching out to the Mississippi River ensured that this would hardly be an inactive partnership. In fact, as has been previously seen, the Northwest Indian War had almost led to direct conflict between Washington and London. Following their loss at Fallen Timbers and with many growing tired of decades of bloodshed, notable Indian leaders, including Buckongahelas of the Lenape, advised accommodation with the Americans. Uneasy peace reigned in the Northwest Territory for a decade.

In May of 1805 an epidemic of either smallpox or influenza ravaged the region, claiming among its victims Buckongahelas. As traditional medicines and practices had failed to save the great leader and many other victims as well, blame was cast upon supernatural actors. An actual witch-hunt began to find those who conspired to either bring the disease upon the tribe or prevent its cure. The leader of these efforts was Tenskwatawa, an orphan who after a series of visions in May of 1805 quickly rose to prominence as the proclaimed Prophet of an Indian revival movement. Tenskwatawa began a series of purification campaigns, discrediting those who sought accommodation with the United States. His followers built a large settlement called Prophetstown near the confluence of the Wabash and Tippecanoe Rivers. The growth and size of the confederation he formed with the help of his brother Tecumseh soon began to worry President Madison and local American leaders. War soon erupted, culminating in General Henry Harrison's victory at Tippecanoe in 1811 and the flight of the Prophet. It was widely believed at the time that not only did the British harbor the Native fighters but provided them with most of the weapons that they used to massacre civilians and battle the US Army. Various voices from General Andrew Jackson to Governor William Blount railed against the English and helped to push the War Hawks in Congress to action.

The same connection can be observed when one examines the second major cause of the war: impressment. The substantial need for sailors to man the Royal Navy during the Napoleonic Wars led Great Britain to adopt the policy of impressing sailors; forcing men to work on His Majesty's ships. Yet the conditions of the vessels and naval warfare during the age led to a problem of mass desertion by impressed sailors. Disease almost certainly played a role in this as well. Sailors, especially those acquired from the prisons of the British Isles, would bring various pestilences aboard the ships which would quickly spread on board the less than sanitary vessels. It is estimated that 5,000 sailors a year from 1793–1815 died aboard British ships.[25] Mass desertions due to conditions led the British to adopt a policy of stopping American merchantmen, searching the crew, and pressganging men to fill their crew lists. As appalling as the practice was to American ideas of neutrality and natural rights, the policy decidedly hit home in June of 1807, when the British warship *Leopard* attacked and seized four men from the USS *Chesapeake* while she was in American territorial waters. As President Jefferson himself wrote, "*Never since the Battle of Lexington have I seen this country in such a state of exasperation as at present.*"[26] Again, disease helped to spearhead the actions that led the young United States to war with the United Kingdom.

Many lessons had been learned from the handling of illness during the American Revolution. In fact out of the first six Secretaries of War, three were physicians.[27] Dr. Benjamin Rush had suggested to President Adams in 1798 that the army and medical services of the nation be closely linked, suggesting that the Physician General, "*should be one of the limbs of a commander in chief*," deciding tactical and strategic issues with the commanding officer.[28] Yet though many of these Secretaries of War may have been accomplished surgeons, not all were competent administrators. Secretary William Eustis especially, who served on the eve of the war, was noted as being both lax and ineffective.

Yet lessons from the past and fear of foreign infection had at least led to two notable pieces of legislation before the outbreak of war. In July of 1798 Congress approved "An Act for the Relief of Sick and Disabled Seamen." The bill called for twenty cents to be deducted from the salary of every seaman coming into an American harbor from a foreign port. These funds were to be used to construct marine hospitals and provide funds for the caring of ill or disabled sailors in the port in which they had been collected. Finally the president was granted the power to appoint a hospital director for each port. This move was aimed at ensuring intercontinental trade and preventing the inflow of disease into America.

On March 23, 1812, Dr. James Smith of Baltimore wrote to Congress, proposing a plan to mass inoculate the population against the deadly scourge of smallpox. Smith hoped to employ Edward Jenner's new method of vaccination that had been developed in 1796 and employed cowpox variolation rather than smallpox. The next year, following the outbreak of war against the United Kingdom, and no doubt fearing a repeat of the smallpox issues of the American Revolution, Congress approved the "Act to Encourage Vaccination." The act required federal agents to preserve genuine copies of the vaccine and vested them with the power to distribute them to citizens. This was followed up by a massive campaign to vaccinate all of the soldiers in the war to prevent a reoccurrence of the incidents of the late 1770s. The models for this included General Washington's own vaccination campaign at Valley Forge and the famous Balmis Expedition of 1803–06.

Congress declared war on the United Kingdom in June of 1812. Ominously, President Madison himself was taken ill with bilious fever, possibly dysentery, and confined to his bed for five weeks in June and July at the height of Congress's deliberations regarding the war. The initial commanders of American troops confronting Canada were little more medically fit than their commander-in-chief. General Henry Dearborn was crippled by rheumatism and General William Hull suffered from heart disease and arterial issues. In fact some historians have diagnosed Hull with depressive psychosis, blaming the condition for his disastrous surrender of Detroit later that year.[29]

Despite the appointment of successive physicians as secretaries of war, the medical corps of the army was also in shambles at the outset of the conflict. *"Military hospitals were to be founded by gentlemen, little versed in hospital establishments, for an army."*[30] Lack of defined powers and roles, inadequate supplies, and infighting kept the hospitals from functioning properly. The changing seasons during the onset of the American campaign, when combined with the presence of thousands of men from different regions and background, quickly led to outbreaks of dysentery and diarrhea in the camp at Greenbush, New York. Located a few miles over the river from Albany, the massive encampment served as a staging area for the movement of troops to the Canadian front. Due to this it saw tens of thousands of men, both healthy and sick. Dr. Mann related the issues confronting the encampment as stemming from poor diet, undercooked food, general miasma, geographic origin of the soldiers, the early onset of cold weather, and the lazy and sickly men who composed most of the soldiers who reported to the camp.[31] The combination of climate and war "*tend to the production of those diseases by which men, whose constitutions have not been assimilated to the climate, are always liable to suffer.*"[32] Other officers placed the blame largely upon the inadequate or unhealthy supplies received from the various states. One doctor lamented the flour that arrived in the camp as being able to "*kill the best horse at Sackets Harbor.*"[33] As supplies dwindled, disease increased; desertions soon followed.

Conditions turned serious with the onset of fall as pneumonia erupted in the encampment. Far from being just an inconvenience, the disease struck hard and fast. Termed Sthenic Pneumonia by contemporaries, the condition proved fatal in two to four days. "*The progress of this disease was most rapid. It assumed in a few hours the strongest marked symptoms of dissolution.*"[34] Time-tested practices of employing bleeding, calomel, or opium proved ineffective, and very quickly the disease spread to surrounding communities. "*This sickness is not as has been represented, confined wholly to the army. The soldiers have only shared in a wide spreading and alarming epidemic, pervading a vast extant of country; the morality of which is without example in this part of the state.*"[35] Soon all of the northern states from Lake Erie east reported an epidemic of the disease. Worse, additional conditions including measles and bilious fever erupted in both the cities and camps of the region as well.

By November, one third of the troops camped at Plattsburgh and Champlain were stricken with measles. One regiment stationed along the border is reported to have fallen from a healthy strength of 900 to less than 200 over the course of only two months. At Buffalo it was reported that eight to nine soldiers were dying a day; numbers which necessitated the burying of two to four soldiers in each grave. Even with this practice, by the spring the cemetery stretched across two acres.[36] Both Burlington and Plattsburgh reported 200 deaths from November to February, with thousands more ill. Some cases morphed into what was called spotted fever, with some victims dying in less than six hours. *"A nurse in the hospital, apparently in good health in the morning at 9 o'clock, became deranged; soon after comatose, at 3 pm was a corpse."*[37]

Death reached up to the front lines at Niagara, and the sickness became one of the prime reasons that the American army proved to be so ineffective at invading Canada during the opening years of the war. In late November 1812, Brigadier General Smyth at Buffalo made a cross-border incursion. General Dearborn had wanted Smyth to attack sooner, but measles and dysentery had delayed his departure. His 1,200 men landed at Black Rock, had lunch, and then quickly withdrew. Measles and peripneumonia had reduced his army in half, making advancing against the British untenable and defeating Smyth's prospects of victory.[38] Conditions worsened upon his return, to the point that bodies began to be buried in shallow graves, whose contents became a ready feast for packs of roaming dogs.[39]

The epidemic of 1812 continued well into 1813. When combined with the hypothermia and frostbite that was ravaging the units located along the border, it was a wonder that the Americans under General Dearborn were able to take York (Toronto) in April. After defeating the British detachments in the town, looting various buildings, and burning down a portion of the settlement, the Americans withdrew. Yet the army that sailed back to New York was *"in a very sickly and depressed state."*[40] Dearborn was stricken with both fever and an attack of angina pectoris, delaying the subsequent attack on Fort George that was to follow the assault on York, by almost a fortnight. Dr. Rush at the time wrote to former president John Adams explaining how three of four generals over the age of 60 were crippled with illness at the time. *"Count Saxe says three things are requisite to make a General, Viz Courage, genius, &* health. *Three of our four Generals are above sixty. . . . Each of those generals has been Laid up with Sickness since this Appointment, and two of them nearly in Sight of the enemy."*[41]

Winfield Scott assumed command of the American forces and took Fort George on May 27, 1813, with very few losses to his side. Unfortunately the conditions for the victorious force once inside the fort proved to be no better than those in the camps. Filthy toilets further heightened the diseases already present in the army. Typhus and diarrhea became common among the troops, and by July General Dearborn himself was relieved of command due to his ill health. *"It would have been entirely agreeable to me, if... you had executed your original intention of providing for your health, by exchanging the sickliness of Niagara for some eligible spot.*"[42] President Madison himself had come down with malaria and was in little better shape than most of his commanders.

By August, over a third of the men in the American army in Canada were reported to be on the sick list, including over half of the doctors. "Lake Fever," better known as typhoid, soon emerged, increasing not only the suffering of the men in camp but hampering the activities of the United States Navy on the Great Lakes as well. Master Commandant Oliver Hazard Perry's great victory on Lake Erie, in fact, had almost not occurred. Perry had hoped to attack in mid-August before the British were able to complete the refit of the 490-ton, 19-gun HMS *Detroit*. Unfortunately "Lake Fever" crippled Perry and an additional roughly 30 men per ship, delaying battle until September 10th and allowing for the presence of the *Detroit*.[43]

Disease would luckily ravage the English army as well, producing a stalemate on the Niagara front. Morale sagged within Fort George as the American regular units withdrew and disease continued to stricken the militia. Even without the primitive medical techniques of the time, it would have been difficult to control these outbreaks as one of the few physicians in the fort, Dr. Shumate, had been killed in a duel in September. By the fall, most of the militia began to desert in order to return home for harvest time. As pressure from the British increased, Brigadier General George McClure was compelled to abandon the fort intact and well supplied in December of 1813.

An additional American force had been organized earlier in the year with the intent of striking at Montreal. Under the command of one of the most notorious men in American military history, General James Wilkinson, this army had experienced hardship almost from the beginning.[44] Moving down the St. Lawrence the Americans had lost much of their supplies, provisions, heavy clothing, and medical equipment. Further supplies had to be brought in from Albany, 250 miles away, which produced more strain upon the army. By the fall, one unit of 160 men reported 75 sick, of whom 39 had dysentery, 18 pneumonia, 6 typhoid, and 12 paralysis of the extremities.[45] Wilkinson would finally launch his assault in early November, meeting the British at the Battle of Crysler's Farm. Unfortunately for him, malaria had heavily damaged his army. Wilkinson himself was, "*much indisposed in mind and body,"* having fallen ill as well.[46] His application of whiskey and opium to his condition only further degraded his judgment in the handling of the campaign. In the end, the American army was defeated by a British force a third its size. Both Wilkinson and Maj. Gen. Morgan Lewis, his second in command, were so ill during the battle that they were confined to a ship and unable to give any orders. By December, 2,000 of the men in Wilkinson's camp were reported sick, of whom 10% would perish. The general would eventually ban funeral dirges in an attempt to hide the true nature of the disaster he was facing. The year 1813 ended no better than it had begun, largely thanks to pestilence.

The first few months of 1814 saw disease further reduce both the British and American armies. Out of the 4,800 American soldiers stationed at Fort George, 1,200 were sick, while almost 50% of the 3,000 English confronting them were diseased as well.[47] On the eastern edge of the front, over 3,700 American soldiers were admitted to the hospital at Burlington.

As the last year of the war's campaigning season drew to a close, the British under Gordon Drummond, the Lt. Governor of Upper Canada, launched an assault on the American-held Fort Erie. Though they outnumbered the Americans, dysentery and diarrhea crippled the English attack and helped the Americans to secure the border. Samuel R. Brown writing in 1814 stated that, *"for every soldier killed, 3 die of disease."*[48]

Congress attempted to remedy the medical situation throughout the course of the war. Various hospitals were established in the most modern of fashions. According to Dr. Mann, *"No army was better supplied, with hospital stores, than that on the frontiers."*[49] In June of 1813 Congress recalled Dr. Tilton due to his experience battling disease during the Revolution and made him Physician and Surgeon General of the Army, while at the same time creating the position of Apothecary General to handle supplies. Disease became a further point of contention between the Federalists and Republicans in the government. As "Buffalo Fever" hit Albany over the winter of 1812–13 and 33 Assemblymen were sickened, Republican newspapers covered up the incident while Federalist ones proclaimed it divine retribution.[50] Similar stories abounded across the nation, with the "Centinel" of Burlington Vermont proclaiming . . .

we are happy to have it in our power to state, that the sicknes among the soldiers in this town has abated . . . from the public prints you would be induced to believe that the troops have been intirely destroyed by sickness, disease, etc . . . the fact is they have wanted for nothing . . . while they were sick for a time, health is now perfectly restored.[51]

Overall, disease would again kill more soldiers than bullets or bayonets during the war. It has been estimated that as many as three-quarters of all American and British casualties resulted from pestilence. General G. Izard's pronunciation of mortality rates from disease as *"prodigious"* remains an understatement.[52] In addition, much as in the American Revolution, disease drove and shaped the strategies and campaigns of the war. The various illnesses of the northern border clearly played a major role in denying the Americans acquisition of Canada and allowing for that country's eventual independence.

A few final vignettes of interest concern the action at Baltimore in September of 1814 that resulted in not only the destruction of the White House but the creation of America's national anthem. Admiral Alexander Cochrane's mission was to land a force under General Robert Ross for a quick assault on Washington D.C. The White House, Capitol, and most other public buildings within the city were set afire and destroyed by this advancing British force. The famous story of First Lady Dolly Madison saving various priceless artifacts from the White House is still a debated subject. Regardless of the veracity of the stories surrounding her, the patriotic and determined wife of the president would not have been able to save anything had her first husband not died from yellow fever during the great Philadelphia outbreak of 1793. Shortly after his death she met and married James Madison, carrying her into the White House a decade later. Following the success of his mission against the capital, Cochrane had wanted to remove the entire fleet from Chesapeake Bay in late August to avoid the onset of yellow fever and malaria, preferring instead the pestilence-free harbors of Rhode Island. General Ross convinced the admiral to remain and attempt to assault Baltimore, but the debate between Ross and Cochrane delayed the operation by almost eight hours.[53] The subsequent attack on Fort McHenry would be a failure for Cochrane and the concurrent assault by Ross on North Point would result in his death. More importantly for American culture, Francis Scott Key during this battle made his famous trip out to the British fleet to help free Dr. William Beanes, a local physician who had been captured and imprisoned by the English.

Disease helped to ensure the growth of the nation with the purchase of Louisiana. Yet as has been seen with the failure to take Canada, pestilence also began to define limitations for the nation's expansion under then-current medical beliefs and practices. An evolution in medical understanding and changes in medical practices would be needed before the country could begin to combat contagion and prosper.

CHAPTER EIGHT

MONTEZUMA'S REVENGE
DISEASE AND MANIFEST DESTINY

"The cholera had broken out, and men were dying every hour."
—ULYSSES S. GRANT, 1846

T he election of Jefferson in 1800 began a slow and steady push of America westward towards the Pacific Ocean. The idea of Manifest Destiny, that it was the nation's God given right to possess the continent, drove American politics for a half century, especially during the 1840s and 1850s. Yet the conquest of the West was anything but bloodless or unopposed. A collection of nations and tribes inhabited the area and posed a formidable challenge to the plans of American expansionists. However, thanks to a collection of wars, explorers, dedicated politicians, diplomacy, sheer luck, and the ever-present factor of disease, the West was eventually subdued. In this case disease helped to limit the amount of war that was to occur in the region.

54°40′ OR FLU
Pestilence had devastated the Native tribes along the Atlantic for centuries. While those of the interior and along the Pacific had certainly not fully escaped its grasp, they had not yet experienced the totality of the destruction that it could affect. This would change with the commencement of American trade and migration through the region, a change that would help facilitate its military conquest.

In February of 1829 the *Owhyhee* under the command of Captain John Dominis made landfall in Oregon.[1] The ship had set sail some 192 days before from Boston. Rounding the Cape, it had stopped at the Juan Fernandez Islands to gather fresh fruit in order to prevent scurvy. Somewhere along the trip though, it had acquired an additional cargo as well: malaria.

In October of 1829 a Mr. Jones who sailed aboard the ship was reported as *"dangerously ill he removed him at once to the English fort 27 miles up the river."*[2] His condition would leave him ashore and under the care of a Dr. McLoughlin for almost six months. Also sent ashore with the sailor was the *plasmodium* that causes malaria. As the Columbia River flooded that year, it provided a natural breeding ground for the mosquitos that transmitted the deadly infection. The Multnomah tribe of Sauvie Island at the mouth of the Willamette River was the first to be impacted. Within three weeks the population of the island had been effectively destroyed by malaria. Over the course of the next three years, 80–90% of the 100,000 estimated overall Native inhabitants of the Oregon region were killed as well, with some villages and tribes reporting death tolls above 95%. As natives fled south towards California to escape the contagion, they carried the disease with them. By 1833 it is estimated that an additional 70,000 Californian Natives had accompanied their Oregonian neighbors to the grave.[3]

The reduction of the Native population of the Pacific Northwest altered the settlement of the region by both Americans and Europeans. Following the Convention of 1818, both the US and UK had agreed to joint occupation of the Oregon region. Debates over an eventual division of the land would continue without agreement for the next three decades. Both nations sought control over the Columbia River as it provided the best deep water ports on the Pacific. Thanks in part to the reduction of the Natives in the region, especially along the river, American settlers were able to populate the area much more quickly than their British counterparts. Already by 1834, only a year after the malarial outbreak had ended, a Methodist mission, a saw mill, and grist mill had all been constructed along the Willamette River. The gradual growth of American settlements and interest in the region would eventually allow for the Oregon Treaty of 1846 which extended the 49° border between the US and Canada westward and therefore granted the United States the prized Columbia River. A more populated and determined Native presence would have certainly hindered American occupation and eventual acquisition of Oregon, yet malaria prevented this.

TRAIL OF TEARS AND TYPHUS

Following the War of 1812 the only major populations of Native Americans left east of the Mississippi resided in the southeastern states and territories. These states, backed by then-President Andrew Jackson, began a policy of forced removal of the Indians westward to modern-day Oklahoma. Though disease did not in any way cause the expulsion of the various Native tribes, it stalked them on their journey, further crippling them and adding to their horrors and reducing the military threat posed by them in later decades.

The Choctaw became one of the earliest victims of forced removals to Oklahoma. From 1831 to 1833, over 17,000 were marched westward. Unfortunately their journey coincided with one of the coldest winters of the century. Deep, penetrating Arctic winds chilled the region, producing ice as far south on the Mississippi as New Orleans.[4] As weather conditions worsened so too did the health conditions of both the Natives and the soldiers sent to accompany them. Pneumonia spread rapidly as both whites and Indians froze to death on the journey. By the time of their arrival, the Choctaw had lost between 2,500 and 6,000 individuals. The Choctaw leader Pushmataha had himself succumbed to disease, dying of the croup in 1824 while on a visit to Washington to persuade Andrew Jackson to support the Choctaw and prevent their removal.[5]

The Cherokee would suffer similar horrors. During their relocation from 1836 to 1839, over 14,000 natives would march out of the South for reservations in the West. Yet due to disease 4,000 would never see Indian Territory. One half of these would contract disease and die during the journey, while the other half would become ill and perish while in government built camps.

> I saw the helpless Cherokees arrested and dragged from their homes, and driven at the bayonet point into the stockades . . . the sufferings of the Cherokees were awful. The trail of the exiles was a trail of death. They had to sleep in the wagons and on the ground without fire. And I have known as many as twenty-two of them to die in one night of pneumonia due to ill treatment, cold and exposure . . .[6]

Tribes that resisted suffered even more harshly, becoming the target of Jackson's military. The Seminoles, secure in the forests and swamps of Florida, engaged the government in war for seven years. Yet by 1842 the tribe was defeated. Disease had claimed the lives of almost 20% of the nation and compromised the Seminole's ability to resist. Of those that remained, around 3,800, or 90%, would be transported to Indian Territory to join their already deported cousins. A year after their removal, an additional 20% of the tribe would die from illness as well, permanently crippling the once unconquerable group.

President Martin Van Buren continued the removal policy after he assumed office. In 1838 the Potawatomi of Michigan and Wisconsin were removed from their homeland by the Treaty of Chicago. One band of over 800 was marched from Indiana to Kansas on what became known as the Potawatomi Trail of Death. Over the course of two months an outbreak of typhoid killed almost 10% of the tribe as it moved westward.

Nor were the relocated tribes the only ones to be visited by illness. Traders accidently spread smallpox to the Mandan Indians of the Dakotas in 1837, yet this was not their only encounter with the pestilence. Fifty-five years earlier, in 1782, smallpox had destroyed seven of the nine Mandan villages. The steamboat *St. Peter's,* which was owned by the American Fur Company, visited the region in June of 1837 and the disease that spread from its passengers annihilated the remaining two bands. Chief Four Bears and the vast majority of the tribe were killed. It is estimated that perhaps only 27 to 150 Natives remained by 1839, forcing them to join the neighboring Hidatsa in 1845. The Mandan were not alone in being afflicted by the smallpox. The Hidatsa and Arikara had their populations cut in half, the Pawnee would lose a quarter of their population, and the Cree and Blackfoot would suffer large casualties as well. In all, it has been estimated that close to 17,000 Plains Indians would die in only a few months from the outbreak, allowing for a more manageable conquest by the US military a generation later.

The Trail of Tears was not caused by disease, nor was disease intentionally spread or encouraged by the soldiers. Yet with such a massive movement of people, over such a distance, and with little to no supplies, it was inevitable. The relocation of Indians to Oklahoma permanently destroyed a significant tribal presence east of the Mississippi. The disease that erupted in the process helped to ensure that that presence would never return, or pose a threat from its new homeland.

HOISTED BY THEIR OWN PETARD

Not all tribes withdrew peacefully before the expanding United States. Some Natives resisted and like the Seminoles were forced to contend with the US Army. Though the progress of the military was in almost all cases unavoidable, it was anything but a quick and easy conquest for America. The disease that had proven so deadly to the various tribes of the continent would now prove to be a hindrance to the expanding Americans as well.

One of these notable incidents was in the Black Hawk War of 1832. Though the entire war lasted only three months, it destroyed the Native American presence in the Michigan region and served as a further impetus for the Indian Removal policies of the 1830s. Initial progress by the US Army proved slow however, and in June President Andrew Jackson placed General Winfield Scott in command of an army of 1,000 men who would move by boat across the Great Lakes towards Illinois.

Scott's offensive took place during the Second Cholera Pandemic that was sweeping the nation and the world. It was inevitable that some of the recruits would arrive with the disease and thus spread it through the expedition. Of the four steamers that left Buffalo, two became so infested with the disease that they didn't make it past Fort Gratiot at the entrance to Lake Huron. The ships docked and released their infected battalions, most of whom either died or deserted upon reaching shore. Scott's own ship the *Sheldon Thompson*, reported no infection before reaching Mackinac. Building upon his experiences with illness during the War of 1812, General Scott had done his best to prepare for a possible outbreak among his troops. Thus when the infection did hit, and though the ship's surgeon was confined to his bed with a bottle of wine, Scott was able to use his knowledge to save the ship and some of his men.

Despite his best efforts, however, almost three quarters of Scott's force had died or deserted before he arrived in Chicago. The general proceeded with the few remaining men under his command inland in a forced march to reach the theater of battle. His army spread the disease wherever they advanced, leading to hundreds of more deaths. Luckily though, Black Hawk's band had already been decisively defeated by General Atkinson at both Wisconsin Heights and Bad Axe, driving them from the state and region.

A similar situation unfolded two years later when, in 1834, the Dodge-Leavenworth Expedition was dispatched to negotiate with the southern plains Indians who were harassing traders in the region. General Henry Leavenworth, the namesake of the future fort and prison, and Colonel Henry Dodge departed Fort Gibson in Indian Territory in June of 1834. By the time the First Dragoon Expedition, as it became known, reached the Washita River, over a third of the men were sick. Leavenworth himself fell ill, passing command to Dodge while he remained behind to recover. After entering Cross Timbers in July, Dodge established a hospital and left more sick behind as he continued into Indian country. Leavenworth eventually proceeded to follow Dodge, but succumbed to illness, dying on July 21st. Dodge did manage to negotiate with the Wichita Indians and by August had returned to Fort Gibson. The sick would continue to straggle in over the course of weeks. Deaths of four or five a day were reported for days after the army's return. Famed artist George Catlin, who accompanied the expedition, himself fell in and recorded in his diary that 150 had fallen ill and died since June. The mission had been a success, but at a tremendous cost which served to further illustrate the disease-filled dangers of both the West and military campaigning during the century.

THE MEXICAN WAR

No president is more associated with Manifest Destiny than James K. Polk. Over the course of his time in office, America defeated Mexico, acquired an immense amount of land from Texas to California, gained control of Oregon Country, and started the processes that would result in both the California Gold Rush and the Oregon Trail. The impressiveness of his accomplishments is only magnified by the fact that he served but one term as president, dying soon after of cholera.

The Mexican War stands out in its relationship with disease for a variety of reasons, most notably its impact on casualty figures and its shaping of Scott's Mexico City campaign. The medical establishment had advanced considerably under Surgeon Generals Joseph Lovell and Thomas Lawson, who were in charge of the Army's Medical Department from 1818 to 1861. Smallpox vaccines had become mandatory for new recruits, as had physicals and the adoption of better uniforms and pay. Yet these could do little to combat the intense level of destruction wrought by illness during America's first tropical war, a conflict that, as a percentage of those who served, was the nation's bloodiest conflict.

As mentioned above, though the smallpox vaccine mitigated that disease during the war, dysentery and yellow fever proved to be just as deadly in the hot southern climate. In fact over the course of the war, while 1,733 Americans were killed in combat, it is estimated that around 11,550 died from disease, a 7:1 ratio. With only around 79,000 men serving in the conflict this means that the United States experienced a casualty rate of 17%. Finally, if the number of soldiers who suffered permanent debilitating illnesses or who died in the years following the war due to the diseases that they had contracted from the invasion are included, the casualty rate has been argued to be around 30–40% of those engaged.[7] J.J. Oswandel wrote in 1885 that more than half of those who returned from the war sickened, died within a few years.[8] As always, just like in both the Revolution and during the War of 1812, volunteers tended to suffer disease at a higher rate than did the regulars, leading Surgeon General Lawson to demand more adherence to the policy of requiring new recruits to undergo physicals.

The most notable battle of the war occurred at Buena Vista on February 23, 1847. General Zachary Taylor was making good progress into northern Mexico and had recently secured the town of Monterrey. However, by this point his army had largely been drained of manpower in order to begin building the invasion force for Scott's subsequent attack on Veracruz. Thus by early 1847 he was commanding only 4,500 relatively untested men. Santa Anna chose this as the moment to strike, and with 20,000 soldiers rushed north to defeat Taylor and thus delay the landing of Scott. Unfortunately for him, the strenuous and ill-supplied march north reduced his army by almost six thousand men.[9] Ravaged by disease, the Mexicans reached Buena Vista and fought the Americans to a standstill before finally being driven from the field. Disease had given the US Army a slight edge, ensured that Scott's campaign would go through, and produced the election of Zachary Taylor to the presidency a year later.

Disease also impacted the planning and undertaking of the first major amphibious invasion in American military history: Scott's assault on Veracruz. Once again, based upon his experience with disease both in the War of 1812 and the Black Hawk War, Scott was concerned about the dangers of the area. He knew yellow fever to be a seasonal infection, abundant along the coast but relatively absent once one crossed the line of the Sierra Madres. The port city of Veracruz was considered the least infested location

Battle of
Churubusco,
1847.

along the Gulf of Mexico, and was therefore chosen for the initial landing. Yet the affair had to be delicately timed. Still driven by miasmic thinking, Scott reasoned that the cold, northerly winds blowing across the region during the winter would drive away the infection. The landing though, had to take place before the end of January, as the disease seemed to resurface in April and a siege of the city and the march inland were expected to take a few months. The entire operation would have to be quick and decisive in order to avoid disaster.[10]

The Mexicans also understood the additional wall that disease lent to their defense. In fact, Santa Anna's strategy aimed to, *"keep his force well in hand, avoiding any general engagement, and distracting the enemy by various movements, until the sickly season; then to bear down with crushing force upon the main body of the enemy."*[11] Further it was reported that the Mexicans hoped that *"the summer season will fall upon them unexpectedly, with its numerous diseases and epidemics, as perilous to the unecolimated; and thus, without a single shot from the Mexican ranks, they will perish daily by hundreds, both men and beasts, who will not have the strength to resist our climate, and in a short time their regiments will be decimated."*[12]

Unfortunately the logistics of the campaign were poorly implemented by the Army. Scott and his men did not land until the first week in March, leaving them precious little time to take the city and reach the safety of the Sierra Madras. In addition, though yellow fever did not yet strike the gathered American armada, other diseases had. Despite Lawson's push for vaccinations, the entire 2nd Pennsylvania Regiment had to be quarantined on Lobos Island due to a smallpox outbreak. They were not alone, "*The Mississippi volunteers . . .have suffered terribly from sickness, and are said to look miserable. The Louisiana Regiment, too, is said to have suffered much from sickness.*"[13] Scott pushed forward with the siege of Veracruz, considered to be the strongest bastion in the Gulf. The bombardment continued for weeks with little result. Scott began to contemplate a frontal assault, which though expected to cost over 2,000 American lives, would save thousands more from falling victim to fever. Luckily the fort surrendered by March 27 and within a week Scott was already on the move pushing inland towards the capital.

By the time Mexico City fell in September, Scott's army was reduced by a third due to a variety of diseases. Much like the Pilgrims at Plymouth, the Mexicans began to consider these deaths to be proof that God was sweeping away the heathen invaders. Though the Army Medical Department worked hard to help mitigate these deaths, various factors worked against its efforts. Surgeon General Lawson personally traveled with Scott on his march, making observations and performing various medical duties. Yet his absence from Washington crippled the administrative structure of the military's medical establishment. In addition, though the use of diethyl ether as an anesthetic had been demonstrated, Lawson forbade its employment in the war. The Surgeon General was moved by concerns over its volatility and the expense to move its associated equipment, as well as studies that suggested it increased risks of acquiring gangrene or producing hemorrhaging among patients. Surgeons would have to wait a generation before employing anesthesia during surgery.

Overall, America proved to be very lucky during its first war in the tropics. While disease claimed thousands of lives, Scott's knowledge of it as a factor in war led him to plan around its presence. In addition, the celerity with which the campaign was carried out helped to compensate for numerous personality and scientific issues which could have easily resulted in an epidemiological disaster.

OREGON AND CALIFORNIA

President Polk's acquisition of California and its surrounding territories allowed for the continued transfer of the American population westward towards the Pacific. To facilitate this, the famed Oregon Trail came into being, moving tens of thousands of migrants from the East over several decades. The transfer of these ordinary civilians did perhaps more than the various actions of the military to complete the path of Manifest Destiny. Yet, though both Mexico and England had been removed as threats in this area during the previous decade, disease had not been.

While the Trail possessed any numbers of dangers that led to thousands of deaths, none killed more than disease. It is estimated that anywhere from 7,000 to 12,000 perished from illness along the route, representing 60% of total deaths.[14] This amounts to almost six individuals buried per square mile along the route. By far the most deadly of diseases to confront the travelers was cholera. The height of the surge west unfortunately coincided with the worldwide Second Cholera Pandemic, and during these years roughly 5% of all travelers on the Trail would die from the disease.[15] And it was not easy to predict who would fall victim:

> Mrs. Knapp, one of the members of the wagon train, died of cholera, and Mother laid her out. Mother took the cholera. Father didn't know what to do, so he had her drink a cupful of spirits of camphor. The other people thought it would kill her or cure her. It cured her.[16]

Yet the alternative, sailing either through the Strait of Magellan or to the Isthmus of Panama, was often times little better. In 1851 Lt. Ulysses S. Grant sailed with the 4th Infantry from New York to California. While en route as he recalled, *"cholera had broken out, and men were dying every hour."*[17] Grant did his best to secure proper medical facilities and no doubt helped to limit the destruction wrought by the illness. Yet despite his best efforts, *"about one-seventh of those who left New York harbor . . .now lie buried on the Isthmus of Panama or on Flamingo Island in Panama Bay."*[18]

Disease that arose from this movement of people to the west only further weakened Native populations in the area and led to one of the most destructive wars in the region. In 1836 a group of Presbyterians had established a mission near Walla Walla, Washington. Whitman Mission, as it became known, was in the heart of Cayuse territory and its inhabitants aimed to trade with and convert the local population. A decade later, a measles outbreak erupted among the Cayuse, killing half the tribe. The traditional practice of alternating between steam baths and cold water immersion that was still employed by these people undoubtedly helped to doom many of those stricken with smallpox or measles.[19]

The tribes blamed the whites of the mission for the disease, despite the fact that they suffered from it as well. On November 29, 1847, a group of Indians raided the Whitman Mission, killing 14 of its inhabitants. The shock of the attack led to retaliation by the Americans in what became known as the Cayuse War. Most of the remaining members of the tribe were killed, leading the US government to seek treaties with other tribes of the region and hurry the process of transforming the region into a territory and eventually a state.

The California Gold Rush saw similar results to what occurred in Oregon. Cholera and other diseases flooded into the territory with the settlers. It is estimated that over 80,000 natives, perhaps half of the population in the region, died from contagion within a few years of the 49ers arrival. In addition, 20–30% of those miners who reached California would die within a year from illness. Fortunately for these adventurers, the death of so many natives from disease as well would prevent an effective resistance against the sickly and weak miners.

The Gold Rush quickly propelled the territory of California towards statehood. However its desire to enter the Union as a free state made its admittance a point of controversy between the North and South. The issue of equal representation in the Senate was again the sticking point, as well as the enforcement of preexisting fugitive slave laws. Various Southern states threatened secession, and the old lions of the Missouri Compromise, Clay, Calhoun, and Webster, again emerged to craft an acceptable pact. Unfortunately, unlike a generation before, the men could not reach an agreement. While Webster was willing to bend on the issue of slavery, Calhoun stood firmly opposed to what he saw as government encroachment on individual liberties, and Clay's insistence on passing the compromise as an Omnibus Bill proved to be a nonstarter. Finally, although President Taylor was a Southerner, he stood opposed to any deal, wishing only to see California admitted. It seemed as if the Union was doomed.

Secession and civil war were luckily averted by a number of concurrent diseases that struck the leadership of America. Both Calhoun and Clay were stricken at the time with tuberculosis. In fact, the former would die of it in

March, removing his objections from the bill and allowing Southern moderates to seize control of the southern Democratic wing of Congress.[20] Clay's inability to push through the bill when combined with his enfeebled condition from TB, led to his withdrawal from the Senate in August. Most vital to the movement of the stalled compromise, though, was the death of the sitting president, Zachary Taylor, on July 9, 1850. While attending a Fourth of July celebration at the Washington Monument, the president consumed cold milk and cherries. Within a few days he was reported to be extremely ill, and was dead in under a week. The cause of his death has been variously argued to have been cholera or gastroenteritis. Regardless, his opposition to the compromise was now a dead letter.

The new president, Millard Fillmore, was more amenable to the Compromise. As well, Clay's replacement, Sen. Stephen Douglas, was able to push through the bill as a series of compromises, thus piecing together various coalitions to achieve the legislation. Thanks to tuberculosis and gastroenteritis, North and South were placated and civil war was diverted for another decade.

Polk's administration saw the largest expansion of America since the presidency of Thomas Jefferson. Associated with this, however, were the hundreds of thousands of deaths that accompanied Manifest Destiny as it marched across the continent. President Polk himself would not seek reelection, nor would he live long enough to see the long-term effects of his actions. Only 103 days after leaving office, Polk succumbed to cholera on June 15, 1849. He most likely contracted the disease during a goodwill tour of the South to receive the praises of those in favor of his expansionist policies. Yet the popular president did not receive the adulations of large numbers of mourners. Fears of his body spreading cholera if handled or approached led to it being buried quickly and without much fanfare.

WOUNDED KNEE

Disease continued to ravage the various tribes of the West well into the second half of the 19th century. As many had experienced few previous extended interactions with the Spanish or Americans, their tribes were hardhit by the infections brought by more recent migrants. Of particular note was a typhoid epidemic that swept through the Western tribes shortly after the end of the Civil War. The Paiute Walker River Reservation in western

Nevada saw 120 deaths alone due to the disease. In all, perhaps 10% of the tribe's population succumbed to the pestilence. Many Natives shunned the settlement for fear of catching the contagion, yet this led only to further conflicts with settlers in the region.

It was at this point that a young Paiute named Wodziwob experienced a millennial vision. He informed the other tribespeople that the Supreme Being was shortly to come live with the Paiute and other Native peoples. Paradise would thus return to the land, as had been the case before the arrival of the white man. This would only come about, though, if the Paiute began to gather and participate in the Ghost Dance, a communal celebration that would imbibe them with power. To a people decimated by war, and most recently disease, this seemed a viable option.[21]

Though initial results of the dance were not promising, word of it soon spread to other tribes and bands. The son of Wodziwob's assistant Tavibo, named Wovoka, would gain fame in the next few years as a magician and further spread tales of the power of the Ghost Dance. Better known as Jack Wilson, the young shaman preached a message of peace and renewal that again focused around the concept of the Ghost Dance. This time the movement spread even further and deeper within the Native community. Frightened and angry settlers, army officers, and Bureau of Indian Affairs agents began to keep a close eye on the dancers. What followed was a slow and steady progression towards the infamous Wounded Knee Massacre in which well over 100 men, women, and children were murdered.

The massacre at Wounded Knee Creek is traditionally used as the date for the ending of the Native American period in United States history. The Indians as a people had been broken. After four centuries of competition, dwindling resources, war, and especially disease, Native American power in the country had been destroyed. Their downfall had begun with the diseases introduced by the Spanish and was finalized with those contagions that led to the massacre at Wounded Knee. Pestilence, even more than the Winchester rifle, had secured the West for America.

CHAPTER NINE

JOHNNY DYSENTERY AND BILLY TYPHUS
DISEASE AND THE CIVIL WAR

*"The slow dead march of camp-disease is much more to
be dreaded than the rapid double-quick of ball and shell."*
—CHARLES FURMAN, 1861

The Civil War is seen as a major transformative event in American history. The idea of states' rights, federalism, the role of government, slavery, and the issue of natural rights were all decided on hundreds of battlefields across the south and west. The United States that emerged after 1865 was a new and altogether different nation from that of only a few years before. The enormous change wrought by this event also touched upon the issue of disease in America as well. Advances in science and military experience allowed for a modicum of control over illness for the first time in the history of the American military. While the war itself was not as heavily directed by disease as had been the Mexican War or War of 1812, contagion still did prove to be a significant foe on the battlefield and in various campaigns. Soldiers would continue to die more from illness than from bullets or bayonets. But while the Civil War helped to more fully incorporate and unite the states, so too would the war help to turn and concentrate military, scientific, and governmental concerns to fighting and controlling contagion in the country in an unprecedented way.

ROAD TO WAR

The American Civil War did not occur suddenly, but rather was the end result of a long evolution of events. Various elements within the South had been actively pushing for secession since at least the administration of Andrew Jackson, with minor pockets passively hinting at it since even the drafting of the Constitution. The presidency of James Buchanan during the late 1850s only hastened the separation of the nation into two separate camps. Buchanan's four years in office are historically ranked as perhaps the worst of any American president. The various actions and inactions of the president would split the nation asunder by the time he left the White House.

James Buchanan's presidency began on a less than propitious note. A few months after his election, the president-elect proceeded to Washington D.C. for his inauguration, and prior to his swearing in, Buchanan stayed and dined at the National Hotel. Located on Pennsylvania Avenue, between the White House and Capitol, the National was one of the most luxurious establishments in the city. Popular in Southern circles, the hotel became the ideal place for a "doughface" to celebrate his inauguration.

After dining one night, the president-elect became violently ill. This was far from an isolated occurrence as it is reported that over 400 additional guests, including many prestigious Congressmen and assorted Democratic notables became sick as well. Over the next year, 40 of these would die, including Mississippi Congressman John Quitman, a renowned Fire-Eater, and Pennsylvania Congressmen John G. Montgomery and David Robison. Rumors and finger pointing quickly began over the cause of the incident and its related casualties.

Poison was instantly suspected, especially following so quickly upon the death six years previously of President Taylor under equally bizarre circumstances. Two theories became widespread: one blamed radical Northern, Republican, abolitionists, while the other considered the entire occurrence to be a Southern plot. While the thought of crazed abolitionists targeting the duly elected president undoubtedly resonated in the minds of Southern Democrats, it made little sense. The death of Buchanan would have placed John C. Breckinridge in the executive's chair. A well-known Fire-Brand, Congressman from Kentucky, and future Secretary of War in the Confederacy, any action that resulted in his rise to power would have only hurt the abolitionist cause. If anyone gained from a plot to poison and remove Buchanan, it would be Southern Democrats. In fact years later, H.J. Raymond, the editor of the *New York Times*, reported:

At lunch today I had a talk with Mr. Forbes. He said he had very good reasons for saying that the famous disease at the National

Hotel, in Washington, in 1857, from which so many people suffered, was the result of an attempt on the part of Southern disunionists to poison Buchanan in order to bring in Breckinridge as president.[1]

Either way it remains doubtful if the attack that struck down Buchanan and killed so many Congressmen was intentional. Other contemporaries placed the blame on accidental poisoning. In this scenario, arsenic poison was used to kill the rats or mice that certainly inhabited the building as they did many others. The rodents then fell into the hotel's water cistern, thus contaminating it with both arsenic and any number of bacterial pathogens. Still a third and more modern theory focuses on an outbreak of dysentery due to the primitive nature of the National's sewage system. Following the outbreak and its associated bad publicity, the hotel worked for a year to fully clean its building and update its sewage system in time for the next opening of Congress in 1859. With no more outbreaks reported, the upgrades seem to have had the desired effect.

Again what stands out about the situation, apart from the terrible loss of life, remains the what-ifs. Had poison or dysentery struck down President-Elect James Buchanan, how would the next four years have unfolded differently? Though a "Northerner with Southern principles," Buchanan was much more moderate than Breckinridge. Yet he still managed to plunge the country deeper into the divide that would split it asunder in 1860. Would a Fire-Eater in the White House in the late 1850s have worsened the situation or produced a more amiable split?

DISEASE IN THE CIVIL WAR

The Civil War dwarfed all previous American campaigns in terms of size, numbers engaged, and casualties involved. Likewise, the role played by disease increased as well. Though hundreds of thousands more troops would succumb to illness than in all previous American wars combined, in terms of deaths to pestilence as a percentage of those engaged, there was a marked decline in casualties from the wars of the early part of the 19th century. Clearly, the means of dealing with illness had begun to change, producing tangible results that undoubtedly saved hundreds of thousands of lives.

Over 250,000 Americans, from both the Union and Confederacy perished in battle from 1861 to 1865.[2] As immense as the number is, it was dwarfed by the sheer weight of those who perished from disease. Perhaps 500,000 are thought to have died off the battlefield from a host of illnesses. When one considers that only around 3.2 million Northerners and Southerners served in the conflict, the rate of death by disease approaches 15%. Yet as deadly as the various contagions were, the death rate was below what it had been in other wars. The Mexican War had seen a 1:7 ratio of deaths from battle versus disease, while the Spanish American War would see a ratio of 1:5. The Civil War's ratio of 1:2 shows an improvement in the handling of illness by the Army. Had the North and South experienced a similar ratio in deaths to the Mexican War, total casualties would have risen above 1.6 million.

Of all the diseases that beset the soldiers of the Civil War, none killed more than amoebic dysentery and diarrhea. Raw recruits traditionally went through a period of seasoning upon joining the army, much as the early settlers did upon landing in the New World. New foods, new water, and new bacteria all wreaked havoc on their immune systems. Diarrhea became known as the "Tennessee Quickstep" among the Northerners, or simply the "Confederate Disease" among the Southerners due to its frequency. The latter became especially afflicted throughout the war due to their consumption of green corn and other substandard rations. A reported 226,828 Southern soldiers are reported to have been afflicted with diarrhea and dysentery in the first two years of the war alone.[3] Over the course of the entire war, both sides reported over 1.7 million cases of the two afflictions, of which number 44,558 died. Almost six times as many soldiers died from an amoeba over the course of the war than from a bullet at Gettysburg.

A variety of other diseases ravaged the two armies as well. Over 20% of all patients in field hospitals were reported to have been afflicted with malaria. This number included 1.3 million cases in the Union army alone. The disease proved less deadly than dysentery, resulting in around 10,000 casualties. Typhoid on the other hand had one of the highest mortality rates for a disease during the war. Of the over 79,000 cases that are reported, around 29,000, or 37% proved to be fatal. One of the most famous sufferers of the disease was Vice President Andrew Johnson. Having contracted the illness over the winter of 1864–65, Johnson was still quite ill during his inauguration in March of 1865. To steady his condition Johnson consumed three whiskeys in the office of outgoing Vice President Hannibal Hamlin, and afterward made a spectacle of himself on the Senate floor. Outraged Radical Republicans went so far as to draft a resolution to Lincoln demanding his resignation, with some even pushing for impeachment. The rumors of Johnson's alcoholism would follow him for years, impacting his presidency as well, though in reality it was a bacteria to blame. Finally, roughly 8.2% of all Union soldiers reported contracting venereal disease during the war. Of course, this number only represents those who actually reported their condition; one can assume that many more probably went unreported. A brief survey of medical records from the time list 102,893 cases of gonorrhea and 79,589 instances of syphilis.[4] A popular slogan at the time reminded soldiers of, *"A night with Venus, a lifetime with Mercury."* Treatment for all of the above diseases and others had not much advanced from the War of 1812 or even the Revolution. Mercury, bleeding, opium, enemas, camphor, and leeches remained within the physician's arsenal, as did some rather bizarre recommendations such as the application of red hot irons to the anus of those suffering from dysentery.

As with many other aspects of the war, a racial divide existed in disease as well. While traditionally casualty counts among black soldiers have often been held up to suggest a reliance upon these men as cannon fodder, a closer examination of the numbers reveals something quite different. While the US Colored Troops lost 2,751 men to casualties in combat, over 68,000 died from other causes, most frequently disease. Black soldiers succumbed to diseases over 2.5 times as often as white soldiers, due in part to their treatment as unequals in terms of food, supplies, and housing. Yet the most pressing of issues, and the one which helped to generate the most deaths, became lack of equal medical care. A few notable exceptions to this reality did arise, most conspicuously at Alexandria. Concerns over the large number of escaped slaves in the area spreading disease to the American army prompted Brigadier General John P. Slough and local city leaders to open a smallpox hospital solely for the treatment of infected blacks.[5] Overall, black soldiers in the war suffered nine deaths from pestilence for every one lost in combat against the South. Finally, a massive smallpox epidemic erupted in Washington in 1862 and spread across the South at the height of the war. Conservative estimates place the death toll among slaves at over 60,000, though the number was most certainly higher.

Disease stands out as well for its impact on the most notorious prison camp of the war and perhaps of the century, Camp Sumter at Andersonville. Well remembered for its abysmal conditions and criminally high death rate, Andersonville brought into focus the horrors of the treatment of prisoners of war in a way that had not been seen since the prison ships of the Revolution. By 1865 over 45,000 Union prisoners had been confined within the stockades of Camp Sumter, a prison built to hold many less men. By wars end 13,000 of these soldiers would be dead. With only one filthy stream trickling through the camp, poor or rotten food, little to no hygiene, and less than 28 square feet of room per soldier, survival at the prison camp became the toughest engagement for men who had witnessed some of the harshest battles of the Civil War.

In the camp's hospital, a reported 17,875 prisoners were admitted with illness or injury. Of this number 12,451 would never leave. Dysentery would claim around 5,605 lives, while scurvy would strike down another 3,614. Between the months of March and September in 1864, when the camp was at its height, 7,712 men would perish of a variety of diseases, including dysentery, scurvy, malaria, and bronchitis. Extrapolated over the course of those months, this amounts to almost 42 men a day, or one man every 35 minutes. It is little wonder that the camp's commandant, Henry Wirz, became the only officer to be executed following the Civil War. Yet that is not to say that Northern prison camps were bastions of cleanliness. Similar conditions and occurrences to Andersonville, though on a smaller scale, were reported at Camp Rathbun in Elmira, New York and Camp Douglas in Illinois. At Elmira nearly 3,000 men, or 25% of the prisoners, would die, even the camp only operated during the last year of the war. At Camp Douglas, over 2,000 cases of smallpox were reported at the prison, out of which 618 men would die. Disease ravaged the camp from 1862 to 1865, eventually spreading beyond the stockades to Chicago. Over the next five years smallpox sickened thousands in the region, killing hundreds in Chicago alone.

BATTLES AND DISEASE

A factor responsible for two thirds of the fatalities of the Civil War must surely have had impact on the individual battles and larger campaigns as well. When the surface of the war is scratched, this fact quickly becomes apparent. Various engagements from the First Battle of Bull Run in 1861 to the demise of the South in 1864–65 were shaped, though not necessarily always decided, by the presence of pestilence.

The secession of South Carolina in December of 1860 marked the beginning of the Confederacy, and the subsequent firing on Fort Sumter four months later plunged the country into open civil war. The original convention on the issue of secession to be held at Columbia had to be relocated to Charleston due to a smallpox epidemic that was then ravaging the city. Regardless however, a unanimous decision among the delegates was quickly reached, and South Carolina became the first state to leave the Union.

Large-scale, open warfare would not begin between North and South until the First Battle of Bull Run on July 21, 1861. Yet prior to this event, disease was already waging a war against both armies. It is estimated that during the summer of 1861 up to 30% of all soldiers came down with illness. Again this was largely due to the seasoning process of incorporating so many different men from different backgrounds into the army. Charles Furman, a Southern soldier, wrote home calling disease the *"universally destructive cavalry,"* adding that, *"the slow dead march of camp-disease is much more to be dreaded than the rapid double-quick of ball and shell."*[6] His regiment of 900 men lost one perecent of its numbers to battle by autumn, and an astonishing seven percent to disease, with many more rendered unfit for combat. At the same time the 7th Louisiana reported 645 of its 920 men down with disease shortly after they were raised.

The Battle of Bull Run itself was fought between armies who were also busy combating disease. From July to September the Army of the Potomac reported 8,000 cases of measles, almost 15% of the army. Their Confederate opponents under P.G.T. Beauregard fared little better. By the time of the battle, the equivalent of two whole Southern regiments, or 1,700 men, were reported as down with measles, and the army desperately needed reinforcements. A soldier of the 11th Mississippi wrote home describing how, *"for the last ten days we have had an average of 40 cases per day."*[7] One Mississippi regiment alone lost over 204 men to measles.

> About one hundred sick men crowded in a room sixty by one hundred feet in all stages of measles. The poor boys lying on the hard

floor, with only one or two blankets under them, not even straw, and anything they could find for a pillow. Many sick and vomiting, many already showing unmistakable signs of blood poisoning.[8]

Additional rebel armies fared little better, with the 10,000 strong force at Camp Raleigh reporting 4,000 cases of measles. Colonel Jo Shelby in Missouri reported, *"Our men from being so poorly clad, and owing to the excessive duties that they have compelled to perform, are rapidly becoming unfit for service . . . our brigade reports now some 500 sick."*[9] His army of 2,319 soldiers was reduced by cold weather and sickness to just over 1,000 men.

West of the main armies at Manassas, the North and South were engaged in a less well known but equally important campaign in modern-day West Virginia. At the time, the area was still part of the state of Virginia and thus officially within the Confederacy. However, residents of the region were overwhelmingly pro-Union, organizing a convention to oppose secession at Wheeling in May of 1861. To both aid this movement and to pressure the western flank of Virginia, a Union army was dispatched to the area before the Battle of Bull Run under the command of George McClellan. Ranged against him was one of the future heroes of the Confederacy, Robert E. Lee.

Though he had been an accomplished soldier in the Mexican War, General Lee's performance in this campaign was abysmal. Lacking the flair and success that was to crown him over the next four years, the Confederate forces were effectively driven from the area. His defeat allowed the residents of western Virginia to declare their loyalty to the Union in June of 1861 and proclaim the independent state of Kanawha (later West Virginia) in October of that year. For months afterwards Lee was attacked for his apparent ineptitude and general over-cautiousness. A Richmond paper at the time referred to the general as, *"outwitted, outmaneuvered and outgeneraled,"* while others attached to him the epithets of "Granny Lee," "The Great Entrencher," or "The King of Spades."[10] After failing to defeat or contain the Union forces, he was ultimately recalled to Richmond and reassigned to a minor post in South Carolina in disgrace.

However, Lee must be somewhat exonerated from the blame for his loss in the West Virginia Campaign, as he was fighting a much more vicious foe than McClellan at the time: disease. In a letter to his daughter written during the campaign, Lee shared that, "*it rains here all the time . . . there has not been enough sunshine since my arrival to dry my clothes.*"[11] The incessant rain flooded the camps and began to lead to increased pestilence among the soldiers. Lee's final major assault upon Union forces in the region at the Battle of Cheat Mountain ended in disaster, not as much through his mismanagement but through the fact that over half of his soldiers were unfit for action. Following his defeat, he was recalled to Richmond, while the onset of winter would eventually increase disease-related casualties in both armies, and by the spring of 1862 action had moved down to the Shenandoah Valley. This would not be the last battle between Robert E. Lee and disease.

Nor were the Confederate units in West Virginia the only army at the time afflicted by illness. On August 20th Lee's opponent in West Virginia, George McClellan, formed the Army of the Potomac and became Lincoln's next hope for a successful invasion of Virginia. Dubbed the "Napoleon of the present war," by a newspaper article at the time, much was expected of the West Point graduate and former railroad executive. By November, due to arguments between the two, General Winfield Scott was driven into retirement and McClellan was named to replace him as General-in-Chief of the Armies. Yet McClellan's habitual cautiousness combined with a bout with typhoid to paralyze the Army of the Potomac for months and to confine its general to his bed for three weeks. In fact his "legendary over-cautiousness" was in part necessitated by the health of his men. "*Gen. Mc-Clellan is worse today, much worse. The danger of a typhoid fever is unconcealed. His case excites a very general interest . . . so thorough as to provide speculatively, even for his successor.*"[12] The General's reliance on homeopathic doctors perhaps hurt public confidence in him more than his questionable tactics. According to General George Meade, it had, "*very much shaken the opinion of many in his claimed extraordinary judgement.*"[13] During his absence, pressure continued to mount on Lincoln for decisive action and a more evident command structure. To accomplish this, the aggressive Edwin Stanton was named as Secretary of War and much more power over the war became centralized in the White House.

At the same time that the Army of the Potomac was paralyzed in winter quarters, various Southern armies fared little better. The Department of South Carolina, Georgia, and Florida reported 41,539 cases of measles dur-

Ward in the Carver General Hospital.

ing the last few months of 1861. While Confederate troops at Mobile were down 13,668 men to the disease.[14] Overall the first year of the Civil War saw little accomplishment on the battlefield, largely due to illness. The armies lost thousands more men to disease than to fighting and had little to show for it. Though the war would proceed much quicker beginning in 1862, this did not mean that disease no longer would play a role.

A notable case where this was true was at Corinth in mid-1862. Union general Ulysses S. Grant had recently won a costly victory at Shiloh, Tennessee in April of 1862. Due to the fact that he was initially surprised by the Confederates, he was replaced after the battle by Gen. Henry Halleck, whose next target was the vital rail hub of Corinth, Mississippi. Unfortunately, uncertainty over the size of Beauregard's Confederate army in the city, combined with the large losses that the Union had suffered at Shiloh, propelled Halleck to move slowly and methodically towards his target, predicting heavy losses in the assault that was to follow. Luckily for the North, there was already a Fifth Column inside the city waging an incessant war against the rebels. Dysentery was already present inside the cramped Confederate quarters within the town, sapping the strength of the defenders. By May, measles and typhoid had become epidemic as well. Even after the arrival of reinforcements under Earl Van Dorn, Beauregard reported 18,000 men sick with only around 50,000 healthy. Filth soon contaminated the hospital also, with 8 out of 10 amputee patients dying of erysipelas, tetanus, and shock.[15] Having lost more men in Corinth to disease than he had lost

Embalming a dead soldier
during the Civil War.

due to bullets at Shiloh, Beauregard evacuated his men from the city before the siege could effectively begin.

Halleck himself would fare little better after taking the city. By June over one third of the men under his command had succumbed to disease. Half of the 29 Union generals in the theater fell ill as well, including John Pope and Halleck himself, with William T. Sherman falling ill with malaria. Due to his losses, Halleck's offensive ground to a halt, with the general refusing to march further south. Progress slowed in the region until well into the fall, allowing the Confederates to reestablish their presence.

McClellan's much awaited and much anticipated offensive of 1862 suffered a similar fate as well. In March, the Union landed over 121,000 on the Virginia Peninsula and marched to Yorktown, where the British army had been enveloped 80 years before. The General's plan called for an advance up the Peninsula and a direct assault on Richmond, a surgical strike reminiscent of his idol Napoleon. Unfortunately, McClellan's cautiousness combined with stiff Confederate resistance to slow the advance to a crawl. After seven weeks the Army of the Potomac had advanced from Yorktown only to Williamsburg, a distance of a little over 30 miles. Reports of sickness began to increase as the Northerners proceeded further inland. Around 3,400 men were struck with infectious hepatitis alone, most likely from contaminated food. As the Union army moved closer to Richmond its numbers began to fall as battle after battle took its toll. Yet, by the time McClellan had driven to within a dozen miles of the capital he still could field around 100,000 men compared to General Joseph Johnston's 60,000 Confederates. As May drew to a close, Johnston launched a surprise assault against the Army of the Potomac, hoping to split it and defeat it, as he knew Richmond would not withstand a prolonged siege. Luckily for Johnston, McClellan had been stricken with neuralgia, or the reoccurring effects of the malaria that he had contracted while serving in the Mexican War.[16] Confined to a bed, the command structure of the Army of the Potomac broke down. Though both sides would claim victory in the bloody battle, McClellan's advance was stopped. Worse for him, with the wounding of Gen. Johnston during the campaign, a much more aggressive commander was tasked to replace him: none other than the formerly derided "King of Spades," General Robert E. Lee.

General Lee subsequently began the Seven Days Battles, and by July 1 had driven McClellan away from the gates of Richmond to a refuge on the James River. By August the remaining mass of the Union army was withdrawn under personal orders from President Lincoln. Though Lee's movements were inspired, the main reason for the retreat of the Northerners was disease. At the end of Lee's campaign, McClellan's remaining 100,000 were bloodied but still held a strong position by Harrison's Landing on the James, impenetrable to Southern attack. Yet the summer season of sickness had set in, and within weeks over 25% of the Northern army was incapacitated or dying. Against the recommendations of McClellan, the diseased force was brought out by boat and returned to Washington, DC.

In the Western Theater the city of Vicksburg, Mississippi proved to be the key to controlling the Mississippi River. After taking New Orleans, Admiral David Farragut moved his flotilla up the river and demanded the surrender of the fortress city. When he was refused, he began a bombardment of Vicksburg in June and July, but with little result. Farragut next considered the idea of digging a canal across the De Soto Peninsula to bypass the fortified heights of the city. Brig. Gen. Thomas Williams was tasked with employing local labor, slaves, and soldiers to complete the task. Unfortunately, waves of malaria and dysentery so reduced the Union digging force that the operation had to be abandoned after less than a month on July 24th. The fortress of Vicksburg would stand for another year. The city would not fall until July 4, 1863, after a two-month siege by General Ulysses S. Grant. Despite a six-week bombardment consisting of tens of thousands of shells, it was the tight siege that would ultimately spell the doom of General Pemberton's 30,000-man Confederate army within the city. Weeks without food supplies and poor living conditions combined to cause an epidemic of malaria, scurvy, dysentery, and diarrhea. In his memoirs, Grant recorded how the onset of war had led to the abandonment by the local population of such tasks as maintenance of the levees. Soon, *"the whole country was covered with water . . . Malarial fever broke out among the men."*[17] By the end of June, over half the Confederate force, as well as many Union soldiers, were incapacitated, and Pemberton had no choice but to surrender unconditionally to Grant. The Confederacy had been cut in two. Luckily for Grant, *"The hospital arrangements and medical attendance were so perfect, however, that the loss of life was much less than might have been expected."*[18]

As the Anaconda Plan tightened its grip of containment of the South, food and medical supplies began to dwindle, and disease naturally increased. The Army of Northern Virginia that marched into Maryland in September of 1862 did so heavily affected by dysentery. Having marched with little more than green corn for food, the health of the men quickly deteriorated. Generals Lee and Longstreet rode into battle in ambulances, with the former having fallen from his horse, injuring both his hands, and with General Longstreet forced to wear slippers due to a heel injury that had become infected and would not heal. The Southern army at the Battle of Antietam was a brave but weakened force. Despite its fighting tenacity, the army's lack of supplies helped to seal the fate of the first Southern invasion of the North.[19]

Confederate raiders became more and more relied upon for the acquisition of supplies for both Southern armies and Southern cities. Unfortunately, these ships also occasionally imported death. On August 6, 1862, the *Kate* arrived in Wilmington from Nassau in the Bahamas. Hidden aboard was the old port scourge of yellow fever. For three months the disease raged through one of the most important ports left in the South. So many citizens died that bodies either rotted in the street or were committed to a mass grave in Oakdale Cemetery. Death wagons roamed the streets late at night as citizens were too afraid to venture outside of their homes. Some took to wearing "gas tars," patches of tar and peanut oil on their clothing that would allegedly drive off the sickness.[20] Overall 654 men, women, and children would die, around 43% of the 1,500 that were reported to have been infected from a prewar population of about 10,000. Other cities would suffer similar fates due to overcrowding and a lack of medical supplies. Atlanta, for example, whose population had more than doubled to over 20,000 due to increased war production, suffered from a series of outbreaks. Smallpox erupted there in 1862, followed by an epidemic of scarlet fever in 1863, and a reoccurrence of smallpox in 1864. A similar yellow fever outbreak in Charleston in October of 1864 would prevent reinforcements from being dispatched to Wilmington when that city was besieged and ultimately fell four months later. All of these various outbreaks, though, did help to shield the coastal areas of the South from full implementation of the Anaconda Plan. Union military planners quickly realized that unseasoned Northern soldiers would not survive long in the disease-infested port cities and forts of the Confederacy, a fact which helped to slow down the overall progress of the war.

The Gettysburg Campaign presents an even more interesting account of the interactions between disease and history. The campaign began with a push southwards by the Army of the Potomac in order to break the stalemate in Virginia and thus march on Richmond. For a week, the Confederates under Lee engaged the Union army of "Fighting" Joe Hooker around the town of Chancellorsville in a series of desperate encounters. Unfortunately, a quantity of smallpox vaccines contaminated with syphilis had rendered almost 5,000 Confederate soldiers incapacitated during the fighting.[21] With Lee outnumbered by over two to one, this outbreak had severely constrained his fighting ability. Yet the Union army fared little better, with an entire regiment, the 20th Maine, quarantined for similar reasons. In the end the Confederates triumphed, and though casualties between both sides reached 30,000, it was to be the death of one man that would drastically alter Confederate fortunes.

General Thomas "Stonewall" Jackson was arguably Lee's most valuable subordinate commander. Despite some claims that he may have suffered from Asperger's Syndrome, he had managed to become a genius in the field of military strategy.[22] Yet all of his accomplishments and battle acumen could not save him from being inadvertently shot by his own troops on May 2, 1863. While returning to camp from an observational ride with fellow officers, Jackson and his men were fired upon by pickets from the 18th North Carolina after being mistaken for a Northern patrol. Jackson was hit by three bullets in his extremities, fracturing his humerus and damaging his brachial artery. In the midst of the confusion, Jackson was not immediately cared for and was at one point dropped from a stretcher during artillery fire. Eventually sedated for the two-mile ride to a safe location, the wounded general arrived and was examined by doctors around two o'clock in the morning. After it was determined that his left arm had to be amputated, Dr. Hunter McGuire administered chloroform to Jackson and removed his damaged limb. A half hour later, Jackson was awake, drank some coffee, and proceeded to fall asleep until nine o'clock on Sunday morning. The next day he rode on a mattress in an ambulance for the 27-mile journey to the plantation of Thomas C. Chandler, where he would convalesce. Though his wounds had been serious, thousands of soldiers had survived similar or worse, including the amputation of a limb.

Over the next few days, however, Jackson would complain of a severe pain in his right side. Due to his rough handling during his evacuation, doctors attributed the pain to simple internal bruising. Yet over the course of the several days his condition did not improve. The application of wet towels to his abdomen, a standard practice at the time, seems to have had little effect on his condition. Doctors finally diagnosed the general with pleuropneumonia, and proceeded to administer mercury, antinomy, and opium as a means to cure the illness. Unfortunately all attempts to save Stonewall Jackson ultimately failed and he died on May 10th, little more than a week after being shot.

Modern medical science has proposed a number of conditions other than pleuropneumonia that could have led to the death of the great commander. Among the various theories are a pulmonary embolism, blood clot, or any number of inter-abdominal issues associated with the gall bladder, duodenum, or pancreas.[23] Regardless, an illness of some form seems to have cut down one of Lee's best commanders shortly before the vital Gettysburg Campaign. Lee's often quoted cry, *"He has lost his left arm but I have lost my right,"* professes to this fact.

General Lee himself was also hampered by ill health, a condition that has been argued to have affected his performance at Gettysburg. Towards the end of March 1863, as Lee was encamped near Gettysburg, he began to complain of sharp pains in his chest, back, and arms. This was at the same time that he was developing a severe cold. His doctors, Lafayette Guild and

S.M. Bemiss, diagnosed him with an *"inflammation of the heart-sac."* They also ordered him to be moved five miles to a nearby house, which was considered to be more conducive to his recovery than a drafty tent. Since the pains continued and were soon accompanied by a high fever, Lee was confined to his bed for weeks. He would complain to his wife of the doctors, *"tapping me all over like an old steam boiler before condemning it."* His illness continued well into April. On April 3rd, Lee would again write his wife, informing her that, *"I have taken violent cold, either from going in or coming out of a warm house."* This was followed by another message two days later in which he described his illness as, *"some malady which must be dreadful if it resembles its name, but which I have forgotten."*[24] His doctors continued to prescribe rest and quinine while waiting for his recovery. The actual diagnosis of Lee's condition has been much debated in both historical and medical circles. Traditional accounts consider his sudden onset of problems to be the signs of a heart attack.[25] Yet months later as Lee began the Battle of Gettysburg his condition seems to have worsened.

The town of Gettysburg itself was known as much for its peach orchards as for anything else prior to the war. The Confederate troops, most of whom had not consumed fresh produce for months, if not longer, rushed to indulge in the edible treasure. Stomach distress began to run through the ranks of soldiers and officers, incapacitating many, including Lee. Speculation has been made that the diarrhea that Lee experienced due to his consumption of the fruit led to hypomagnesemia.[26] A condition that results in irritability, mental disorientation, mental confusion, and overall weakness, hypomagnesemia well explains the poor generalship and condition of Lee at the critical moments of Gettysburg. The Confederate general was sick, but far too proud to turn over command to Longstreet. A month later, on August 8th, he would write to President Jefferson Davis that, *"I sensibly feel the growing failure of my bodily strength. I have not yet recovered from the attack I experienced the past spring. I am becoming more and more incapable of exertion."*[27] Lee recognized the failure in his health that caused his defeat, and the downturn in Confederate prospects. Peaches and heart disease helped to stop the Southern advance.[28]

Lee's health would similarly impact the Overland Campaign as well. As Grant began his push into Virginia, Lee faced the almost impossible task of holding the Union back in what had become a war of attrition. For weeks

Grant moved south, suffering high losses but inflicting proportionately equal ones upon the Confederacy. One of the few opportunities to contain him would occur at the North Anna River on May 23rd, when Grant mistakenly divided his army only to find Lee holding a commanding central position capable of destroying either wing. Unfortunately, at that juncture Lee had again been stricken with a case of diarrhea so severe that he was confined to his tent for days. Unable to spring his trap, Lee allowed Grant, who had suddenly realized his peril, to continue moving to the southeast, eventually crossing the Pamunkey River to the environs of Richmond. Yet Lee was not the only commander to suffer operational failure due to illness. One of his finest corps commanders, A.P. Hill, was equally stricken, and kept from issuing orders at the Battle of Spotsylvania from May 8–21. Having contracted an STD during his West Point years (possibly gonorrhea), Gen. Hill would occasionally experience reoccurrences of the illness.[29] Some of these relapses, such as at Spotsylvania, proved severe enough to effect his performance as a battlefield commander.

The ravages of disease were not confined to the human species during the Civil War. Just as harmful for Confederate attempts to slow down the Union advance in the east was an outbreak of glanders that ravaged the South's horses. Hints of the disease were already appearing in 1862, with Confederate Secretary of War George W. Randolph being notified by Lee that *"disease among the horses"* was limiting his mounted troops.[30] Epidemic from December of 1863 until the end of the war, this illness ultimately severely reduced the ability of the Confederate cavalry to hold off its Northern adversary. The ill-effect was magnified by the Southern practice of having cavalrymen supply their own mounts, whereas the Cavalry Bureau in Washington could scour the entire North for fresh horses. Thus, disdaining the normal procedure of slaughtering sickened animals to prevent the spread of glanders, General Lee opted to wait out the illness in the hopes of having as many mounts recover as possible. Undoubtedly this merely helped to spread the pestilence more quickly.[31] Due to the equine disease, as well as the increasing battle experience of Union horse soldiers, the Southern cavalry arm grew comparatively weaker as the war entered its final stage. The glanders outbreak would also serve to severely limit the agricultural recovery of the South following the war, serving as a lasting reminder of the Rebellion.

One of the more notorious disease-related incidents of the war centered around Kentucky congressman and future governor Luke P. Blackburn. Interested in medicine from an early age, young Blackburn had helped his uncle tend to the ill during the great cholera epidemic of a decade before. After attending medical school he made a name for himself by successfully utilizing quarantine during the 1853 yellow fever outbreak in the Mississippi region. With the beginning of the Civil War, Dr. Blackburn cast his lot with the Confederacy. Too old to fight, he offered his services to the medical department but was turned down. Eager to play a role in the conflict, he removed himself to neutral Canada, a gathering place for many Southern agents and sympathizers, and helped to procure goods for Confederate raiders. Yet this less than glorious contribution to the war was not enough for the Kentucky physician.

While daydreaming of adventurous ways to aid the Southern rebellion by damaging the North's ability to wage war, Blackburn read of a yellow fever outbreak in Bermuda. Determined to act, in April of 1864 he proceeded by ship to the island in order to "aid" the victims of the contagion. Over the course of two trips to the infected colony, Dr. Blackburn proceeded to collect clothing, bedding, and bandages contaminated with the blood, feces, and vomit of yellow fever victims. According to a disgruntled partner in his plot, the Confederates' plan involved sending trunks containing the disease-ridden fabric to various major Northern cities, including one addressed personally to President Lincoln. The plan apparently never reached fruition, though even if it had, since yellow fever is transmitted by mosquitos and not contact with the clothing of the infected, it would have produced no infection anyway. Yet Blackburn's suspicious actions, when combined with the testimony of his former partner, proved enough to lead to his arrest in Montreal at the end of the war. Charged with violating Canada's neutrality law, his trial made headlines in America, catching the attention of Sec. of State William Seward:

> He's under the charge of felony, in this that he conceived and put into execution . . . a plot to disseminate contagion and pestilence in this and other cities of the United States, by clandestinely transmitting for an unsuspicious market here masses of infected clothing taken from the corpses of persons who had died of yellow fever

in the tropics. It is not easy to understand how an offense of that character, which is a detestable crime against mankind, can be supposed, even by the felon himself, to be entitled to be regarded as an act insurrection, rebellion, or civil war. The president's proclamation offers no immunity in this case.[32]

Blackburn would eventually be acquitted, but would stay in exile in Canada until the 1870s. In 1873 he returned to the United States, venturing to Memphis, Tennessee to help fight a yellow fever outbreak, for which he received a hero's welcome. Four years later he performed similar actions in Florida, and in 1878 helped treat pestilence sufferers in western Kentucky. His popularity led to his election as governor of Kentucky in 1879, in which position he served until 1883. Yet his attempt at the first use of biological warfare in modern American history earns him a special place in the history of the nation, one ironically absent from his tombstone, inscribed simply with "The Good Samaritan."

MEDICAL DEPARTMENT

The Union Surgeon General William A. Hammond once famously said that the Civil War, *"was fought at the end of the medical Middle Ages."* Indeed the innovations produced in the field of military medicine during the war represented an evolution equal to or beyond the advancements that occurred in any other area during those years. The Medical Department of the Army in 1861 was largely still a product of the Mexican War. Even the Surgeon General, Thomas Lawson, at 71 years of age was the same officer who had headed the medical department since the 1830s. Continuity in this case meant stagnation. If the medical establishment was to function in an effective way during the war, it needed to modernize.

Surgeon General Lawson would pass away from apoplexy in May of 1861, a month after the outbreak of hostilities. Leadership in the medical department would prove to be an issue during the first few years of the war much as it had been during the Revolution. Surgeon Robert C. Wood, the natural successor to Lawson, was passed over due to his familial connections with President Jefferson Davis. Lincoln instead chose the experienced Clement Finley to replace him. Yet he would not last a full year in his position due to personal conflicts with Secretary Stanton. Against the latter's advice, Lincoln next chose the young William Alexander Hammond to fill the role, a position that he would hold until August of 1864.

The medical establishment was so unready for the war that various private organizations arose to fill the void. Chief among these were the US Sanitary Commission and the Western Sanitary Commission. The former was organized by a group of concerned women in New York City, while the latter started in St. Louis. Both worked towards raising money, acquiring volunteers to help care for wounded and sick soldiers, and pushing legislation to benefit the fighting man. As part of this, it was the U.S.S.C. that convinced Congress in April of 1862 to reorganize its Medical Department and begin a series of reforms to aid in the treatment of the sick.

The various Surgeon Generals instituted a series of far reaching reforms. Among these were an increase in the number of medical inspectors, an army medical museum, an army chemical laboratory, an army medical school, and a permanent army medical hospital in the capital. Likewise, in 1862, Hammond constructed Satterlee Hospital in West Philadelphia. At the time it was the second largest medical facility in the country and could house over 4,500 men. By May of 1864 it had treated over 12,000 men while suffering only a few hundred fatalities. Smaller field hospitals were also constructed to service the dead and dying on battlefields. Yet the Seven Days Battles demonstrated that the army lacked a system by which to move the wounded or sick to local or distant hospitals to provide aid. Jonathan Letterman worked with Hammond to adopt a system of ambulances to transport men from the field. In place by Antietam, the US Army was able to save countless thousands from a premature death, evacuating every dead or injured man from the battlefield by nightfall. On the water a similar situation unfolded with the commissioning of the first hospital ship, the USS *Red Rover*. Initially used to quarantine smallpox victims, the ship eventually evolved into a floating surgical platform in the Western Theater.

Yet not all of Hammond's reforms were popular, particularly his handling of various diseases. Early in the war the Surgeon General had pushed the use of an oral prophylactic of quinine to help prevent malaria, rather than simply battling it once it had arisen. In a similar vein in May of 1863 Hammond banned the use of mercury calomel for the treatment of illness, due to his concerns that it could lead to mercury poisoning.[33] For generations, doctors had prescribed "*opium for looseness, calomel for tightness,*" thus

Hammond's actions were seen as blasphemy to many physicians in the Army. Especially with his preference for no action rather than using mercury, a near riot broke out in the medical department, known as the Calomel Rebellion. This proved to be the opportunity that Stanton needed to rid himself of the adversarial Surgeon General. In September of 1863, Hammond was sent on a tour of the occupied South, following which a rigged court-martial removed him from command. History would vindicate his medical decisions, and generations of future soldiers and civilians owe their lives to the ideas and strategies of Hammond.

The Civil War transformed the country in a variety of ways, both positive and negative. At the same time, disease continued to impact the way that battles were fought and that strategies unfolded. Yet the war also presents a change in this narrative. The various undertakings by the government and individuals sought and served to lessen the overall damage wrought by disease upon the nation. The lessons learned and the impetus acquired would gather speed over the next century as man slowly gained the upper hand in the millennia-long war against pestilence.

CHAPTER TEN

REMEMBER THE MAINE, TO HELL WITH YELLOW FEVER
IMPERIALISM AND ILLNESS

"Is M. de Lesseps a canal digger, or a grave digger? Men
die and are quietly put in a hole along the track. Dozens
of the laborers have died already and dozens more are sick."
—THOMAS NAST, 1881

A s the American frontier closed, the eyes of many turned overseas to satisfy their economic, social, or political ambitions in keeping with Frederick J. Turner's Frontier Thesis. These experiences and the advent of the nation's brief flirtation with imperialism would drastically alter the course of American history. Much of the Progressivism and global interventionism that would following during the 20th century can trace its roots to imperialism. Yet gaining land in regions renowned for tropical diseases that had killed thousands of American and British soldiers during the wars of the 18th century would have been a pointless endeavor had not equal advancements occurred in the nation with regards to disease management. Following the basic thesis of the interaction between disease and civilization, the growth of one provoked an advance in the other and vice-versa.

INDIAN WARS
The various engagements fought by the US Army against the remaining Natives out west were often punctuated by outbreaks of contagion. In fact the various diseases that had ravaged the area over the last half-century helped to cripple the fighting ability of many of the holdout tribes. Yet even with this added Fifth Column, the troopers of the US Cavalry still faced numerous difficulties in their various wars. Occasionally, in fact, disease

172

hampered their advance as much as it did the defense of the Indians.

Much as with the cavalry of the Confederacy in the 1860s, the outbreak of a horse disease in the 1870s paralyzed one of the most effective elements of the military. The Equine Influenza Epidemic of 1872, which originated in Markham, Ontario, swept across the United States during the fall. It eventually became one of the catalysts behind the Panic of 1873 and the subsequent Long Depression, the most devastating financial collapse of the century. America's reliance on horses for many aspects of society and the economy magnified the effects of the epizootic disease many times over.

The disease was first reported in Canada in September of 1872, and by October had reached New England. Before the month ended, trade and travel in the state of New York was brought to a standstill. November saw the pestilence reaching south to Florida and it was reported throughout the West by December. The entire nation's economy began to succumb, and though railroads continued to run for a while, the lack of horses to pull coal or move goods ultimately brought them to a halt as well. A lack of horses to pull fire engines in Boston crippled that city during a great conflagration on November 9th, which resulted in 13 deaths, the loss of 776 buildings, and the destruction of $75 million worth of property. At the same time, US Cavalry forces fighting the Yavapai and Apache in the West, were forced to campaign without mounts or supply trains due to the epidemic. The infamous Skeleton Cave Massacre of the Yavapai came about from a band of Indians forced to fight guerrilla style rather than in cavalry raids. Though 90% of all horses would recover by the next year, the effects on the economy would be felt for years to come: a major, though non-human epidemic had crippled the economy.

SPANISH-AMERICAN WAR

Save for various engagements against the Indians of the West, the American people had largely been at peace since the end of the Civil War. As the 19th century came to a close, however, a variety of catalysts would propel the country into war with Spain. At a little over three months, the war was one of the nation's shortest conflicts. In addition, with only around 300 deaths in combat, the war was one of America's least destructive major operations. Yet disease once again influenced the fighting of the campaign. More importantly, the investigation and treatment of the various pestilences encountered in the Caribbean would have a lasting impact on American world power for the next century.

War had been raging in Cuba for three years before America's subsequent intervention. During this time perhaps the greatest ally that the Cuban rebels had was yellow fever. Though they had revolted many times before against the Spanish, most notably from 1868 to 1880, a combination of Spanish brutality and military might, a lack of foreign support, and dissensions within Cuban society had doomed all attempts at independence. An influx of American economic investment, the rise of local labor unions, and the actions of Jose Marti restarted the drive for freedom in 1895. Lacking supplies, weapons, and holding only small portions of the island, the rebels relied on sheer tenacity to drive out the Spanish. Yet their most potent weapon quickly became yellow fever.

The disease had been rampant on the island since at least the 17th century. In 1649 alone, "*a third part of the population was devoured from May to October by a species of putrid fever, which carried off those attacked in three days.*"[1] From 1853 to 1879 the disease would continue to strike almost monthly, carrying off around 56,000 in those 26 years.[2] The Spanish imperial army fared little better than the local population, losing around 16,000 men from 1895–98. By the time America became involved in the war it is estimated that only 55,000 out of 230,000 Spaniards were healthy enough to fight. Dr. Santiago Barroeta, who fled Cuba during the war, reported at the end of May that, "*it is something awful to see. Those ignorant, sickly peasants brought here from Spain to defend the Spanish flag are dying by hundreds every day. The epidemic is so acute that many die in two or three hours.*"[3] Due in part to this prob-

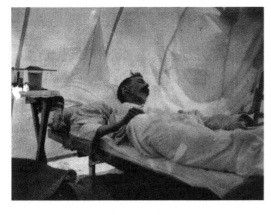

Yellow Fever victim.

lem the Spanish instead resorted to brutal tactics and concentration camps under the infamous "Butcher" Weyler. The estimated third of the island's rural population that died due to his actions did much to push the United States into the conflict.

Though America's invasion of the island was brief and relatively painless, the presence of yellow fever was known, feared, and consequential. At the time, one of the reasons even advanced for the war itself was that the conquest of Cuba and cleaning up of Havana would eradicate yellow fever in America.[4] Building upon not only the experiences of the US military during the various tropical wars of the earlier 19th century, but its recent urban bouts with the illness as well, the government expressed concern regarding the timing of the invasion and the tempo of the advance across the island. In fact the presence of the disease affected even the building of an army. President McKinley was advised shortly after war was officially declared to request from the states an army of 200,000 as opposed to his proposed 100,000, due to the possibility that many would fall victim to fever.[5] Secretary of War Russell Alger himself asked Congress in late April for six regiments of "immunes," which Senator Donelson Caffery of Louisiana stated could be raised from New Orleans alone with relative ease. Senator Joseph Hawley of Connecticut introduced a bill two weeks later to recruit 10,000 such men from the Southern states.

Due to the presence of illness, Secretary Alger's larger strategy for winning the war involved relying heavily upon the US Navy to sink the Spanish fleet and secure the islands of Cuba and Puerto Rico, followed by a quick, concentrated invasion. These troops would then be as quickly with-

Disease related deaths at Spanish concentration camps in Cuba.

drawn once their objective was complete. *"I do not think the fatalities growing out of yellow fever need be so great as to stop us from entering Cuba whenever it shall appear necessary to bring the war to an end."*[6] Former Surgeon General John B. Hamilton, who had done much to revolutionize quarantine procedures, argued that as the filth of certain cities and towns of Cuba was to blame for the various fevers of the island, the spreading of contagion could be controlled, much as the British had done in Jamaica.[7]

Keeping in line with the above way of thinking, the campaign commenced with the Surgeon General of the Navy, Dr. Van Reypen, ordering the complete epidemiological mapping of the waters of Cuba. Areas along the coast or in harbors where sewage entered the ocean or garbage was dumped were labeled as potential febrile locations. Among these, Santiago stood out as one of the worst, where *"dead cats, dogs, etc, are thrown into the gutters, where they remain with the offal and surface sewage to pollute the atmosphere."*[8]

Yet despite all precautions and warnings, the invasion of Cuba still took place at the height of summer. On June 22, 1898, the first American troops landed at Daiquiri relatively unopposed and quickly moved inland. However, only two weeks later, on July 6th, the first American soldiers would fall victim to yellow fever at Siboney. As the first units began to entrench around Santiago, the main objective of the campaign, the death toll in the rear began to increase. In response, General Nelson Miles ordered on July 11th for the village of Siboney to be burned to the ground to clear out the contagion. A variety of illnesses soon began to conspire to halt the attack against Santiago.

General William Rufus Shafter's campaign against Santiago was almost a disaster for a variety of reasons. The general himself landed in Cuba at over 63 years of age, weighing over 300 lbs. and suffering from severe gout. His preexisting condition, along with new health issues that arose on the island, combined to keep him at the back of the army during much of the fighting. In fact during the disembarking of troops at Daiquiri, Shafter would remain shipboard leaving the march inland to the aged former Confederate, General Joseph Wheeler. The latter soon proceeded to disregard orders and develop his own campaign. Even after the landing of Shafter, his inability to receive detailed information on the fighting at the front plus his judgment being impaired due to his health caused the army to became bogged down. As fever crippled other officers, changes in leadership began to occur, including the rise to power of Colonel Theodore Roosevelt. *"On the afternoon of the 25th we moved on a couple of miles, and camped in a marshy open spot close to a beautiful stream. Here we lay for several days . . . General Young was struck down with the fever, and Wood took charge of the brigade. This left me in command of the regiment."*[9] The battle that was to follow at San Juan Hill was to propel the young, former Assistant Secretary of the Navy to the presidency, a situation only made possible by the illness that upset the pre-established command structure.

By July 17th Santiago had fallen, with Spanish General Jose Toral realizing his dire situation of complete encirclement and with his men dying daily from disease. Yet the damage to the American units around the city had only just begun. A little over a week later, 293 men in Shafter's own headquarters were stricken with yellow fever. Almost 3,800 additional troops in the army came down with illness, of which 2,900 were fever related, or close to 20% of Shafter's overall command. Commanders blamed the soldiers' inability to boil water, their bivouacking outside, and the failure of the Army to demolish disease-infested homes. It soon became apparent that should the Americans stay in Santiago they would be wiped out in a matter of weeks. Colonel Roosevelt would write years later that had they stayed, barely *"half the men would live through to the cool season."*[10] General Shafter himself once claimed that disease was a *"thousand times harder to stand up against than the missiles of the enemy."*[11]

As authorities in Washington dragged their feet, the various commanders in the field took an unprecedented step. An open letter from the American generals in Cuba was sent to Senator Henry Cabot Lodge and leaked to the press as well, expressing the frustration in the field with the inactivity in Washington. *"If we are kept here it will in all human possibility mean an appalling disaster, for the surgeons here estimate that over half the army, if kept here during the sickly season, will die."*[12] The reaction from the American public was palpable, tarnishing the McKinley administration's perceived fine handling of the affair up to this point. Roosevelt's participation in this incident could possibly have contributed to the stonewalling of his Medal of Honor award. Despite fears of returning yellow fever back to the continental United States, the War Department announced on July 28th that all American combat troops would be evacuating the island. Secretary Alger returned healthy soldiers directly home while the sick were quarantined at Siboney.

Operating under the still prevalent belief that black soldiers were immune to yellow fever, the 24th Colored Regiment volunteered to serve as nurses to the sick American soldiers. Unfortunately one third of the 460 men in the regiment would fall victim to the illness within only 40 days. By the time peace had been formalized a few weeks later, less than 350 total soldiers had died in battle, while over 2,900 had fallen from disease.

Most of these deaths would actually occur stateside in the various camps set up for recruits during the war. Surgeon General George Sternberg had issued his Circular No. 1 to specifically address the issue of sanitation and illness in camps. Concerned specifically with typhoid, Sternberg advised that troops without access to approved water sources boil their fluids. In addition, he advised that camps be located far from stagnant water and constructed with proper drainage. Finally, waste materials were to be removed and latrines dug and moved often. Though the Surgeon General's orders reflected a miasma view of contagion, they undoubtedly did contribute to the reduction of some of the germ-carrying agents normally present in traditional military camps.

Yet despite these precautions, four months into the war typhoid had reached a higher epidemic proportion than during the same timeframe of the Civil War. Of the over 171,000 troops camped in the Southeast, almost 21,000 contracted the illness and were incapacitated. Sadly, almost 1,600 of this number would eventually die. Overall, typhoid represented 82% of all sickness stateside and cost the US Army almost 20 regiments of potential soldiers. In total the disease accounted for 87% of all deaths in the war. A soldier was 5.6 times as likely to die from disease as from a bullet or bayonet.

Matters in the field were made worse due to a beef scandal that arose out of Chicago industries selling tainted and adulterated meat to the troops in Cuba. Numerous gastrointestinal issues and illnesses sickened and killed an unknown number of troops, especially those already suffering from malaria or yellow fever. The issue of *embalmed beef,* as General Miles famously called it, only further tarnished the image of Secretary Alger and further weakened the American soldier in the field. To help prevent similar issues in the future and calm the country before the November elections, President McKinley established a commission under General Grenville Dodge. The Dodge Commission largely identified longstanding organiza-

Spanish American War nurses.

tional issues within the military for the disease-related problems that arose in Cuba and denied General Miles' accusations about tainted beef. Though no individuals were identified in the Commission's final report, Secretary Alger's days were numbered, and President McKinley forced him to resign in July of 1899.

Besides introducing America to imperialism, the war produced many important results in the field of epidemiology as well. The national disgrace of the typhoid deaths on American soil in 1898 prompted Surgeon General Sternberg to set up a Typhoid Board under the command of Dr. Walter Reed. Perhaps the greatest contribution of the group was its discovery that the illness was spread by flies and humans contaminated by fecal bacilli.[13] Hundreds of years of miasma theory had suddenly been altered. Dr. Reed then went on to help in the clean-up of Cuba during that island's occupation by America. Much good was done with the construction of hospitals and the reduction of epidemic diseases, particularly dysentery and typhoid. The US Army also updated its curriculum at West Point to include classes on field sanitation and hygiene, as well as organizing an Army Nurse Corps and a Medical Reserve Corps.

PANAMA CANAL

Perhaps the single most important outgrowth of the American invasion of Cuba was the continued medical research of Dr. Walter Reed into the vectoring agent of yellow fever. Though Carlos Finlay, a Cuban doctor, had theorized in 1881 that mosquitos were the carriers of yellow fever, his theory, much like those of Pasteur attacking miasma, were largely ignored for a generation. The US Army finally established a board in 1900 to study tropical diseases, under the aegis of Dr. Reed. The famed physician hoped to utilize human volunteers to test various theories of transmission, including by air, mosquitos, or physical contact. Members of the board, including Jesse William Lazear and Clara Maass, infected themselves to expose the cause of the deadly disease. Members of the public were also invited to participate and were well compensated with between $8,000 and $20,000 in gold payments.

Numerous members of the experiment would die attempting to discover the vectoring agent. Famed nurse Clara Maass survived her first bout with the illness, only to die after being re-infected in an attempt to test whether or not infection brought immunity. The American public at the time certainly understood the great sacrifice that had been made to obtain such priceless information. The names of those who died were added to the US Army Roll of Honor, and in 1928 a Broadway musical debuted entitled *Yellow Jacket*, which depicted the heroics of the experiment.

Yellow fever had been a scourge of America for centuries and had served as an effective barrier to its Caribbean expansion. Most recently an 1855 outbreak in Norfolk, a city of 16,000 people, had resulted in the flight of 6,000 and the sickening of almost all of those who remained. The steamer *Ben Franklin*, sailing from St. Thomas to New York, had put in for repairs and brought the disease to the port city. A twenty-foot-high wall was constructed to quarantine sections of Norfolk, but angry residents later burned both it and the local hospital. Though over 500 barrels of lime were blanketed over the city, little respite was felt until the first frost. In the end over 3,000 people, almost 20% of the town was killed. A similar outbreak in New Orleans in 1853 resulted in the deaths of 8,000. Shreveport would bury 759 of its 4,500 citizens in 1873 in the famed Yellow Fever Mound. Finally, famous political cartoonist Thomas Nast would himself succumb to the plague while serving in Ecuador in 1902.

One of the most influential and costly outbreaks of the disease occurred in Panama in the 1880s. Based upon their success in Egypt a generation before, the French under Ferdinand de Lesseps embarked on a project to construct a canal across Panama in 1881. The project, which was estimated to take 12 years to complete, would instead be abandoned in only eight. Conditions in the country, when combined with malaria and yellow fever, destroyed the French army of excavation. Disease claimed around 60 lives during the first year of work, over 120 the next, and over 420 in 1883. Casualties mounted exponentially as the digging slowly advanced, eventually resulting in an estimated 22,000 deaths, a similar number to what the French had suffered during the Siege of Paris by the Prussians less than 20 years before. The thousands of lives lost and over $287,000,000 spent on the project proved to be a financial and humanitarian disaster for the country.

Not surprisingly, when President Theodore Roosevelt seized upon the chance to restart construction in 1904, many feared a similar outcome. Yet for both military and economic necessity, Roosevelt felt that the project was vital to the nation. This fact was brought home to the American people following the epic and time-consuming journey of the USS *Oregon* around South America during the Spanish-American War. Two years into construction, the United States had already spent hundreds of millions of dollars and reports showed that over 85% of canal workers had been hospitalized. Doctor William Gorgas, who had used the theories of Walter Reed to eradicate yellow fever in Havana, was placed in charge of sanitation in Panama in 1904. Unfortunately his $1,000,000 price tag and controversial theories garnered little support on Capitol Hill. However, as the death toll mounted and with upwards of a quarter of the workforce having fled, President Roosevelt decided to act, granting Gorgas everything he needed. Beginning in 1905 the doctor and his 4,000-man sanitation team began an all-out assault on the mosquito population of Panama. Stagnant water was drained, polluted areas were cleaned, and oil and pesticides were sprayed across the isthmus. In the end the team consumed 120 tons of pyrethrum, 300 tons of sulfur, and 600,000 gallons of oil. In addition, Gorgas employed mosquito netting, more effective systems of quarantine, and improved the sanitation of the various towns and cities of the region. Free medical care was provided to all workers, including the provisioning of one ton of prophylactic quinine each year to prevent malaria.

The work of Reed and Gorgas allowed for the construction of the Panama Canal. Yellow fever was effectively eradicated by 1906 and malaria was brought under control a few years later. Though thousands of Americans would still die from accidents and disease, they would not see the same tallies that afflicted the French. The canal would be completed in time for World War I and would secure American trade and naval domination for the next century. All of this resulted directly from the work of Walter Reed and the various diseases of the Spanish-American War.

THE PHILIPPINE INSURRECTION

Despite the apparent popularity of the Spanish-American War, the prospect of the nation engaging in imperialism wasn't heralded by all as the proper course of action for the United States to take. Very quickly after the acquisition of Puerto Rico and the Philippines an anti-imperialist organization arose within the country, supported by some of its most prominent citizens. The vicious fighting that characterized the Philippine Insurrection following America's acquisition of that colony from Spain served to turn more people away from the path that McKinley and Roosevelt were leading the nation down. As destructive as the guerrilla war was, though, disease proved to be a more deadly opponent than the Filipinos, as well as a de facto ally for the Anti-Imperialist League.

American operations on the islands began almost immediately after the occupation of Manila and the issuance of the proclamation detailing annexation. Unfortunately for the Filipinos, the outbreak of war was accompanied by a resurgence of the old nemesis of bubonic plague. This epidemic would ravage the islands, beginning with Luzon, from 1899 until 1903. Overall 600 would die from the contagion with thousands more sickened. Worse for the archipelago, though, was the introduction of cholera from China as part of the 6th Pandemic of the late 19th century and early 20th centuries. Produce imported from Hong Kong is historically blamed for the illness, but it was the conditions set in place by the military occupation that turned the outbreak into a disaster. Cholera in fact had largely avoided the Philippines during previous outbreaks as the sparse settlements of the islands discouraged its spread. The counterinsurgency strategy employed by General MacArthur and others countered this historical demographic protection.

The cholera outbreak would have remained confined to Manila had it not been for the guerrilla actions of the Filipino fighters. In response, the American military established concentration camps, a model used with varying degrees of effectiveness by the British in South Africa and the Spanish in Cuba during the previous decade. The impact of confining tens of thousands of villagers into close quarters naturally led to the outbreak of disease, and in this case to cholera gaining a foothold on the islands. In addition, the release of these people following the declaration of the end of hostilities helped to push the bacteria to the far corners of the Philippines. The pestilence attacked in two waves, the first spanning from 1902–03 and the second from mid-1903 to 1904. Death rates are estimated to have been 31 per 1,000 for Americans on the islands and 108 per 1,000 for Filipinos. Overall 4,000 died in Manila alone and estimates for the entire nation range from 100,000 to 200,000, ten times the casualties from fighting. While both the American military and government attempted to set up quarantine camps in 1903 and 1904, recent memories of the concentration camps kept many Filipinos from actively using them.

Besides sowing death in the Philippines, disease also helped to fuel the anti-imperialism movement back home. The first Americans to land on the island came as garrison troops, and as such were often quartered with and around locals. Not only did fraternization with local women become commonplace, an apparent flood of prostitutes from around East Asia and from as far away as America and Italy descended upon the islands. This *"cosmopolitan harlotry"* quickly led to an epidemic of sexually transmitted diseases.[14] Worse for the Filipinos, the soldiers disseminated the various contagions as they moved throughout the countryside engaging the rebels. Conditions worsened to the point that by 1900 over one-sixth of all men in the First Reserve Hospital in Manila (almost 3,000 in total) had been admitted due to STDs. Stricken soldiers soon took to jokingly referring to themselves as *"Rough Riders,"* a maudlin attempt at associating martial honor with their unfortunate conquest.[15] Though the Army responded by setting up inspection centers to identify women infected with *"Asiatic diseases,"* and then deport them, and even introduced boxing into the nation to distract soldiers and keep them abstinent, members of the Anti-Imperialist League took a different approach.[16]

During the era of Social Darwinism and national eugenics, the miscegenation of the white race with that of the Filipinos and their venereal diseases became a national concern. Members of the Anti-Imperialist League lamented the moral weakening of the current race and the genetic degradation of its future members through both alcohol and sexually transmitted diseases. Bessie Scovell of the Women's Christian Temperance Union wrote at the time that, *"Again and again has my blood boiled at the hundreds of American saloons being established throughout our new possessions. And, shame of shames, our military authorities in the Philippines have introduced the open and official sanction of prostitution!*"[17] Edward Atkinson, one of the founders of the League, attempted to send pamphlets to every soldier in the field discussing the dangers of STDs, but was restricted from doing so by the Postmaster General.

In the end, 4,234 of the roughly 100,000 American men who served in the campaign died, the majority again from disease. In fact in 1900 alone, 70% of all deaths in the US Army were from illness.[18] The average soldier experienced four illnesses each year while serving in the Far East, costing the army strength and the taxpayer money.[19] Overall the war continued to serve to show that warfare frequently spread disease. At the same time, however, the effects of disease not just on the individual but upon society helped to usher in the national surge in Progressive thought that was to dominate the next decade.

CHAPTER ELEVEN

LOVE IN THE AGE OF CHOLERA, WARFARE IN THE AGE OF TYPHOID
PROGRESSIVISM AND PESTILENCE

*"Ultramodern war, twentieth century war,
the war of the scientist and the laboratory."*
—JACK LONDON, 1910

The advent of the 20th century brought a new concept into American political thought: Progressivism. This form of positivism asserted that the careful scientific study of a problem and the application of precise governmental power to solve it could be used to improve the overall condition of the majority of Americans. While the idea was most notably applied to cleaning up factories and slums, improving education, and providing for consumer protection, the conquest of contagion was also a concern for Progressives. Yet in a far darker application of the practice, some began to

WWI US Army poster.

suggest that not only could disease be eliminated from society and war, but could also be employed to the benefit of the nation against its enemies as well.

DISEASE AND WORLD WAR I
The Great War of 1914–18 can truly be considered the first modern military conflict. Science and strategy were combined by the opposing sides in an

attempt to gain victory in a way that had previously been unconsidered. While the evolution of airpower, chemical warfare, submarines, and tanks are well known and appreciated, the role of disease has been less studied. Yet building upon centuries of interactions between disease and war, and recent efforts by progressive governments to control illness, it is not surprising to find both the Allies and Central Powers attempting to harness contagion as a weapon.

Famed American author Jack London wrote a short story in 1909 entitled, "Yah, Yah, Yah." The tale revolves around an aged Scotsman named McAllister who ruled over Oolong Atoll in Melanesia. Yet the natives of the islands hated him, constantly wishing death upon their foreign despot . . .

> But McAllister lived on. His health was superb. He never caught fever; nor coughs nor colds; dysentery passed him by; and the malignant ulcers and vile skin diseases that attack blacks and whites alike in that climate never fastened upon him. He must have been so saturated with alcohol as to defy the lodgment of germs. I used to imagine them falling to the ground in showers of microscopic cinders as fast as they entered his whiskey-sodden aura. No one loved him, not even germs.[1]

The narrator wonders why the Natives don't rise up against McAllister, being that there are 5,000 of them and only one of him. He is told a tale by one of the islanders of how years before, the savage attacks that the inhabitants committed upon trading vessels had led to an assault by a small fleet on Oolong Atoll. After defeating the aboriginals, the whites put ashore six captured Melanesians, each stricken with measles. Over half of the population of the atoll would die, forever convincing the locals to leave white men, even overbearing despotic ones, alone.

London's tale was more than just a flight of fantasy by a writer. To many of the era, biological warfare was modernization. If disease could be mastered to save one's countrymen, then it could be used to defeat one's enemies. The same author would go on to spell out his thought more succinctly in the short story "The Unparalleled Invasion." This futuristic tale depicts a rising China which modernizes in the 1920s, eventually overwhelming Japan and becoming the dominant power in East Asia. However, unlike Japan, it does not seek to expand militarily, but instead relies upon the "*fecundity of her loins*," to eventually out-reproduce and slowly over-whelm the rest of the world. All seemed lost until Jacobus Laningdale, an official at the New York Health Office, recommended a radical proposal.

> But on May 1, 1976, had the reader been in the imperial city of Peking, with its then population of eleven millions, he would have witnessed a curious sight. He would have seen the streets filled with the chattering yellow populace, every queued head tilted back, every slant eye turned skyward. And high up in the blue he would have beheld a tiny dot of black, which, because of its orderly evo-lutions, he would have identified as an airship. From this airship, as it curved its flight back and forth over the city, fell missiles—strange, harmless missiles, tubes of fragile glass that shattered into thousands of fragments on the streets and house- tops. But there was nothing deadly about these tubes of glass. Nothing happened. There were no explosions. It is true, three Chinese were killed by the tubes dropping on their heads from so enormous a height; but what were three Chinese against an excess birth rate of twenty mil-lions? One tube struck perpendicularly in a fish-pond in a garden and was not broken. It was dragged ashore by the master of the house. He did not dare to open it, but, accompanied by his friends, and surrounded by an ever-increasing crowd, he carried the mys-terious tube to the magistrate of the district. The latter was a brave man. With all eyes upon him, he shattered the tube with a blow from his brass-bowled pipe. Nothing happened. Of those who were very near, one or two thought they saw some mosquitoes fly out. That was all. The crowd set up a great laugh and dispersed.[2]

The various nations of the globe had joined together and launched a biological attack of every known pathogen against the Chinese. Hundreds of millions would die and in the end the nation and race would fall. Euro-pean and American leaders would divide up the lands of the Chinese empire and the world would move on. London finishes his tale by asserting that though germ warfare had become the new standard, the various na-tions of the planet realized its destructiveness and held a conference at which they banned its future use. Twentieth-century war had become *"the war of the scientist and the laboratory."*

The outbreak of World War I would bring many of London's thoughts to fruition. Chemical agents such as chlorine, phosgene, and mustard gas led to the deaths of an estimated 90,000 on the Western Front, with over a million others affected. Yet attempts to use disease have been shrouded in secret and rumor. From a variety of governmental reports, scientific studies, and the often dubious journalistic work of Wickham Steed, we can begin to piece together a history of the biological program of Germany during the war. The Central Powers seem to have had the most active of the biological programs, or at least the most documented and revealed to the public.

The main German biological focus appears to have been a campaign to employ anthrax and *Burkholderia mallei*, or glanders, to infect the horse population of various Allied and neutral nations. The role of the horse both on the battlefield and in industrial life was still prominent enough to make it of strategic concern to both sides. In fact, at the start of the war in 1914 cavalry still accounted for almost one third of most European armies. Undoubtedly, German military planners appreciated the impact that glanders had produced in the Army of Northern Virginia in 1864, and in numerous other wars of the late 19th century.

As a potential future belligerent and a current supplier of vital supplies to the Allies, the United States became a prime target for just such German sabotage efforts. Numerous physical attacks took place on American soil from 1914–17, most notably against Black Tom Island in New Jersey and the US Navy Yard in California. Biological attacks began just as early, with the Germans attempting to ship horses and sheep already infected with anthrax and glanders to America and other nations in order to start an epidemic. Unfortunately this proved to be more difficult, more liable to be noticed, and less effective than beginning an outbreak directly within the enemy nation. For this to happen, German agents would need to bring the bacteria to America, culture it in secret, and then spread it to its intended targets. Thus the German government began the process of infiltration into the country of German scientists and terrorists.

One of the first confirmed attempts occurred early in the war and involved a naval officer named Erich von Steinmetz. After sneaking into

America, allegedly disguised as a woman and possessing vials of glanders and other bacterial weapons, Steinmetz was tasked with infecting local livestock. Unfortunately for the Germans, by the time the vials were brought to a lab to test the samples, the organisms inside were already dead. Undeterred, Berlin simply redoubled its efforts. George W. Merck, president of Merck & Co. and one of the heads of the American biological weapons program in World War II, wrote in 1946 that there was, *"incontrovertible evidence . . . that in 1915 German agents inoculated horses and cattle leaving United States ports for shipment to the Allies with disease producing bacteria."*[3]

In order to aid their strategy, the Germans went so far as to set up a biological weapons laboratory in Chevy Chase, Maryland, around 1915. Their agent in this attempt was Dr. Anton Casimir Dilger, a natural born American and son of a Medal of Honor winner. Dilger had been educated in medicine in Germany and was in that nation at the start of the war. Using his American passport, he was able to freely travel between the countries, returning to the United States in 1915 with vials of anthrax and glanders. Along with his brother Carl, Anton Dilger set up a laboratory in Chevy Chase just outside of Washington DC. From here bacterial samples were sent to Baltimore and as far away as St. Louis, in which city Dilger attempted to establish a second production facility. Anthrax was sent to fellow German agent Capt. Frederick Hinsch, who lived in a house in Baltimore. Hinsch would go on to disseminate some of his weapons to German agents in New York City. Some of these saboteurs even infiltrated the horse stables in Van Cortland Park in order to infect the animals there. One of the agents was eventually captured, reporting that . . .

> The germs were given to me by Captain Hinsch in glass bottles about an inch and a half or two inches long, and three-quarters of an inch in diameter, with a cork stopper. The bottles were usually contained in a round wooden box with a lid that screwed on the top. There was cotton in the top and bottom to protect the bottles from breaking. A piece of steel in the form of a needle with a sharp point was stuck in the underside of the cork, and the steel needle extended down in the liquid where the germs were. We used rubber gloves and would put the germs in the horses by pulling out the stopper and jabbing the horses with the sharp point of the

needle that had been down among the germs. We did a good bit of work by walking along the fences that enclosed the horses and jabbing them when they would come up along the fence or lean where we could get at them. We also spread the germs sometimes on their food and in the water that they were drinking. . . . Captain Hinsch spoke often when I met him of different fires that had occurred and of outbreaks of disease among horses and would make remarks about how well things were going.[4]

Yet this promising campaign begun by the Germans would amount to little. While it was reported that additional numbers of livestock were also infected in Norfolk, Newport News, Baltimore, St. Louis, and Covington, it is difficult to trace all of these to German action. Regardless, with the onset of winter most of Dilger's bacterial samples died off. By 1916 he himself traveled to Germany, perhaps to obtain more bacterial and viral agents. Upon his return to America he became aware that the FBI was quickly closing in on him, forcing him to flee to Mexico, and from there to Spain.[5] Later Dilger became one of the prime participants in the German scheme to bring Mexico into the war against the United States that resulted in the infamous Zimmerman Telegram. Though no massive bacterial or viral outbreak was to erupt in America or cripple its role as a military supplier to the Allies, the fears and lessons of German actions would prove invaluable during World War II and the Cold War.[6]

In addition, the actions of Dilger and others led to fear of domestic terrorism. Various unproven or unfounded stories began to circulate of terrorist actions by various foreign or domestic actors against America. One of the most infamous tales involved infected plasters in 1917. In July of that year, a story began to circulate in local newspapers that a German immigrant was distributing poisoned bandages in both Kansas and Illinois. Various reports claimed tetanus, pneumonia, typhoid, or spinal meningitis to be the infections that would potentially devastate the region. Bauer & Black Company in Chicago quickly denied the story, causing the possibility "*ridiculous*." The Department of Justice issued warnings to the public against using the product, which simply fueled the fears and rumor mills then circulating across the nation.[7]

The German Empire soon expanded its biological efforts to the rest of

WWI medical transport.

the world as well. On the Eastern Front, attempts were made to infect sheep in Romania bound for export to Russia, hoping to cripple the food supply. Germany was accused of dropping chocolates and toys infected with disease for children in Bucharest.[8] Rumors even began to circulate that Berlin had attempted to unleash the plague in St. Petersburg in 1915.[9] Agents were in fact arrested attempting to sneak into both Russia and Romania in 1916. Most notable among these was Baron Otto Karl von Rosen, a Norwegian aristocrat who was captured while attempting to infect horses in his home country with anthrax. In addition, the conquest of Serbia was quickened by a massive outbreak of typhus which decimated the nation's fighting age population.

On the Western Front, Germany reportedly poisoned French wells in 1917 using the corpses of deceased soldiers.[10] A year later, during the general German retreat, efforts were made to release cholera and glanders upon the advancing Allied forces. Yet again all of these attempts produced no discernible deaths and many may have been baseless accusations reminiscent of similar assertions made against Jews in the Middle Ages, or claims made regarding the Rape of Belgium. Perhaps the only success of the German biological warfare program was the infection of 4,500 mules in Mesopotamia, dramatically crippling British supply lines in the theater, though not changing the outcome of the campaign.

American biological efforts focused more on prevention than on offensive capabilities. Surgeon General Rupert Blue headed the nation's health service from 1912 to 1920 and during his tenure and that of his predecessor the military began to mandate more vaccines for its soldiers. As part of this, from 1904 to 1913, 585,000 soldiers and sailors were vaccinated against tetanus. During that same nine-year period the armed forces would report only eight cases of the illness, demonstrating the effectiveness of the program.[11] With the outbreak of war, Blue ordered the mass production and stockpiling of vaccinations for tetanus, smallpox, diphtheria, and typhoid by the Hygienic Laboratory. All of these represented preventable diseases that historically caused the most deaths in war. In an effort to reduce the impact of disease, President Wilson in 1917 ordered the Public Health Service to be placed under the umbrella of the US Army, where it would stay until 1921.

A much more far-reaching endeavor of the Surgeon General and the Public Health Service was the campaign against venereal disease. An appreciation of the effect of venereal disease upon an army combined with a progressive concern for society led the military and government to seek to curtail the institution. In fact, over 13% of all recent draftees were found to be stricken with syphilis during their initial medical evaluation. The Bureau of Social Hygiene had been formed in 1913 under the watchful eye of John D. Rockefeller, Jr. in order to assault prostitution due to its disease-related effects. After the outbreak of World War I, Congress moved to address the issue with the Chamberlain-Kahn Act of 1918. This legislation gave power to the armed forces to indefinitely detain individuals who were deemed dangerous to the military. Over 20,000 women would be arrested and quarantined over the course of the war. They were forcibly held, examined, and imprisoned, some for up to a year.[12] In the end, many who did not suffer from venereal disease would acquire it due to the forced examinations. Yet the efforts of the government were largely successful, as prostitution was pushed further towards extinction and the nation as a whole would suffer no recorded casualties from these diseases during the war. In fact 96% of the 48,167 cases of venereal disease that were treated by the military were contracted by the men before they joined the Army.[13]

The one offensive biological program by the United States that received attention involved the weaponization of ricin. A toxin derived from the castor oil plant, ricin is extremely deadly if injected or inhaled. Attempts were made by the US from 1914 to 1918 to either deliver ricin by bullets and shrapnel or in a cloud-burst form. However the former method was problematic, as ricin was not heat resistant, and the latter was considered impractical until an antitoxin could be mass produced. Due to this, neither process was ever perfected and the research came to an end by the conclusion of the war.

Overall casualty figures for the American military during World War I showed a vast improvement from previous campaigns. Official tallies list 116,516 men as having been killed during the conflict. Of these 46%, or around 53,000, were deaths in battle, while the remaining 54%, or around 63,000, were from disease or accidents. Though the modern medical knowledge and tactics of the American Army had reduced the number of deaths due to illness that had plagued it in Cuba, Mexico, and even Canada, the notion of losing one soldier from typhoid for every one from a bullet was still unacceptable. The disease that caused the overwhelming number of casualties for the military was pneumonia. Between 73–84% of all deaths from illness resulted from that one disease, with all others claiming less than 5%.[14] In retrospect, we know that these high numbers of secondary pneumonia occurring in young, healthy males was actually the result of a much deadlier enemy: Spanish Influenza.

SPANISH INFLUENZA

As World War I was coming to a close, a far greater calamity was about to befall the planet: the Spanish Influenza Pandemic. Over the course of only a few months in 1918 and 1919, over 30,000,000 around the planet would die of contagion, with perhaps 48% of the world's population infected.[15]

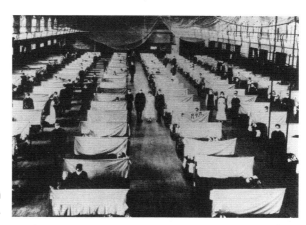

Spanish Influenza makeshift hospital.

Though the disease would disappear as quickly as it had arrived, it would send more people to the grave in a short amount of time than any other illness in history. Created by the conditions of the Great War, it would be spread by the global nature of the conflict and would bring the horrors of the war to almost every town and home in America.

The Spanish Influenza struck the world in a series of three waves. The first generally acknowledged outbreak of the flu occurred in March of 1918 in Kansas. A group of men from Haskell County, Kansas who had signed up for the armed forces was reporting to Fort Riley for basic training. Eighteen of the recruits were reported as sick with symptoms of influenza to the Public Health Service at the time of their arrival at the base. By the second week in March, the fort reported over 500 cases of the disease, and it only spread from there. The soldiers departing for Europe brought the disease with them to the continent and outbreaks were reported there by May. Yet due to wartime censorship few reports of the actual extent of the illness showed up in newspapers. Only from Spain, which was neutral during the conflict, did stories emerge of the true horrors of the epidemic. An estimated 80% of the population of that nation contracted the illness, including King Alfonso XIII. The lack of censorship in Spain and frequent stories of the King's illness and recovery led many to believe that the disease had originated there.

The illness seemed to strike young adults the hardest. Soldiers in the prime of their life and in perfect health were struck down in a matter of days. It appears that the disease led to an increased risk of the development of pneumonia, which brought about the majority of deaths. This was caused by a cytokine storm, a condition in which a body's own immune system overreacts to an infection. Lungs can become filled with fluid, and tissues and organs would be severely damaged. Death followed quickly, with nurses and doctors able to do little to aid the sick or dying . . .

> We didn't have the time to treat them. We didn't take temperatures; we didn't even have time to take blood pressure. We would give them a little hot whisky toddy; that's about all we had time to do. They would have terrific nosebleeds with it. Sometimes the blood would just shoot across the room. You had to get out of the way or someone's nose would bleed all over you.[16]

The illness soon spread to Europe and quickly moved through the ranks of the soldiers in the field. By May of 1918 over 10,000 sailors of the Royal Navy were stricken with influenza. The Grand Fleet became unable to leave port, and the German High Command began to plan in order to take advantage of the situation. Unfortunately, Spanish Influenza reached the continent soon afterwards and struck soldiers on both sides. The ongoing German Spring Offensive launched in March soon ground to halt in part due to the incapacitation of so many units. General Erich Ludendorff, the mastermind of the final two years of the war, was himself said to have suffered from mental illness and to have sustained a serious breakdown in 1918 that possibly impacted the planning and execution of the offensive.

After a mutation, or antigenic shift, the disease reemerged at various points on the globe in August of 1918, including in Boston. The first cases and deaths were reported at both the naval facility in that city and Camp Devens in Middlesex County. Within the next few weeks, 5–10% of the dockworkers on Commonwealth Pier were killed. Overall, 2,000 would become victims in Boston before the contagion had burned itself out. By September the Spanish Influenza had hit divergent parts of the nation including Texas and California. Philadelphia became one of the hardest hit cities in the country after the disease was first reported among the men in the naval yards on September 18th. By the next day, 600 sailors were stricken ill. Deaths would continue for months, with 1,700 dying in one 24-hour period alone. The quarter million citizens who participated in the fourth liberty bond drive parade through 23 blocks of Philadelphia on Sept. 28th merely marched the contagion into the heart of the city. Philadelphia's board of health began a war on spitting, sneezing, and coughing, and as the fall wore on, required all residents to wear gauze masks if out in public.

Yet no preparations could alleviate the onslaught once it began. Due to the war the availability of nurses proved to be insufficient. Students and volunteers began to fill the role but never in enough numbers. The Philadelphia Automotive Club volunteered to transport nurses around the city to help victims of the flu. Yet the movement of bodies became a larger problem. Corpses piled up in homes, warehouses, and funeral homes. Wagons and trucks roamed the streets in ways reminiscent of the Black Plague, calling out for residents to turn out their dead. Mass graves were dug by highway crews with promises to unbury them for proper funerals once the storm had passed. Various cities began to close down factories, schools, churches, and movie theaters. Though Harvard University remained open, it mandated physical fitness and hygiene classes for incoming freshmen. Garbage went uncollected in the streets and telegraphs went silent as their operators either died or stayed home.

The Public Health Service did not begin accepting reports of the contagion until September 27th, but by then it was too late. The PHS was itself undermanned and underfunded due to the war, and had to seek an immediate grant of $1,000,000 from Congress. Masks and alcohol seemed to be the only means of combating the illness. In addition, wartime censorship restricted the dissemination of news and tactics. In October alone, 195,000 Americans died of the plague, the worst month of the outbreak. Only cold weather halted the disease, and though a third wave would sweep across the planet in the spring of 1919, its fury was spent.

By the spring, 675,000 Americans had died with an estimated quarter of the nation having been infected. An additional 43,000 United States soldiers succumbed to the disease in Western Europe, an amount nearly equal to the number that had died from battle wounds in the war. The average lifespan of the nation dropped by 10 years, with entire towns being depopulated.[17] The disease also cost hundreds of millions in lost wages and trade, and had the war continued another year would have severely hampered the physical and industrial capacity of America to fight. Yet despite this it was largely forgotten in collective memory for a variety of reasons.

THE HEALTH OF PRESIDENT WILSON

A more permanent effect of the Spanish Influenza may be its impact upon the negotiations taking place at the Versailles Peace Conference. For four months President Woodrow Wilson had negotiated and fought with Prime Minister David Lloyd George of the United Kingdom and Prime Minister Georges Clemenceau of France. Though he had arrived in France preaching the Fourteen Points and promising no harsh punishments for the Central Powers, Clemenceau, especially, stood in Wilson's way. After months of impasse, Wilson even threatened to return home to America, rather than compromise on his ideals. Yet this abruptly changed on April 3rd at around 6 p.m. President Wilson suddenly became violently sick, with high fever, coughing, and diarrhea, and was stricken to his bed. The suddenness of the attack led some of his closest associates to assume that he had been poisoned. Horrifyingly, one of the president's aides who had also taken ill would die only four days later. Though not known for sure at the time, the cause of these sicknesses seems to have been Spanish Influenza.

In the midst of his illness, Wilson called upon George and Clemenceau to meet with him to continue talks. Yet various aides and political officers noticed a marked change in the president. He was quick to anger, forgetful, delirious, and unable to think in the sharp manner in which he could before the attack.[18] Prime Minister George himself recorded Wilson's *"nervous and spiritual breakdown in the middle of the Conference."* Numerous other members of the American delegation would resign in protest as the President suddenly began to give in to French, Japanese, and Italian demands. The harsh punishments demanded by Clemenceau and enacted upon Germany due to Wilson's sudden change of heart would help to create the conditions that would fuel the subsequent rise of the Nazis and lead directly to World War II. Yet had Wilson died, the result at the time was feared as being even worse. According to Wilson's private doctor, Cary T. Grayson:

> The president was suddenly taken violently sick with the influenza at a time when the whole of civilization seemed to be in the balance. And without him and his guidance Europe would certainly have turned to Bolshevism and anarchy. From your side of the water you can not realize on what thin ice European civilization has been skating. I just wish you could spend a day with me behind the scenes here. Some day perhaps I may be able to tell the world what a close call we had.[19]

Unfortunately, a worse condition was to arise four months later. Perhaps due to his historically poor health or an after-effect of the influenza, President Wilson suffered a massive stroke on October 2, 1919. His wife Edith found him prone and bleeding on the bathroom floor of the White House, and instantly called for Dr. Grayson. The President had experienced at least three previous strokes and had just completed a month-long coast-to-coast tour of the nation that took him across 8,000 miles and to over 40 cities in an attempt to sell the League of Nations to a skeptical American public. The strain appears to have been simply too much for his perennially weak constitution.

At such a historically important juncture, Edith Wilson and Dr. Grayson assumed the best course of action would be to launch a massive cover-up of the President's illness. His exact condition was kept secret from all, including his own vice president and cabinet. With the left side of his body paralyzed, his vision impaired, and his cognitive functions compromised, the situation would almost have certainly demanded his resignation or removal. Wilson was kept confined to his bedroom, with Edith serving as his official go-between for the next year and a half; in fact especially for the first few months, she effectively ran the executive branch and nation. Though word did begin to leak out almost immediately, the full extent of his illness was unknown. Former President Taft wrote only three days after Wilson's stroke that Secretary of the Treasury *"McAdoo says the President is in a state of collapse—that his mind is clear but that he is so weak that his doctors would not permit him to discuss or think about any of these matters. . . . He says that he would like to help, but he is in a delicate situation, being the son-in-law of the President."*[20] Regardless, for Wilson and his ideals a worse damage resulted. In March of 1920 the Senate voted down the Treaty of Versailles and America never joined the League of Nations, effectively crippling that organization. Between the President's bout of influenza and his stroke, the path had been laid out for the slow march to World War II.

CHAPTER TWELVE

BULLETS, BAYONETS, AND BOTULISM
BIOLOGICAL WARFARE IN THE TWENTIETH CENTURY

*"But old War was made a thing of laughter. Naught
remained to him but patrol duty. China had laughed at
war, and war she was getting, but it was ultra-modern
war, twentieth century war, the war of the scientist and
the laboratory. Hundred-ton guns were toys compared
with the micro- organic projectiles hurled from the
laboratories, the messengers of death, the destroying angels
that stalked through the empire of a billion souls."*
—JACK LONDON, 1910

Though disease had been only sparingly used as a weapon during the Great War, military planners had at least awakened to the idea and terror of biological warfare. Indeed, a brief perusal of history serves to justify their hopes for harnessing the power of illness to defeat one's enemies. If a nation could employ the strategy of the Mongols at Caffa, Cortez at Tenochtitlan, or the British at Fort Pitt on a larger scale using modern science and technology, it could cripple its enemy with little to no collateral damage to itself and leave valuable infrastructure unaffected.[1] More humanely, disease does not always bear the same level of fatalities as carpet-bombing, or nuclear warfare. A nation can be crippled by smallpox or typhoid and still see a vast majority of its citizens recover and survive, with relatively little to no long-term damage. Though the Geneva Convention of 1925 specifically forbade the use of such weapons, it did not prohibit the production or movement of them. Thus once war seemed imminent, all sides began to mass produce and stockpile biological and chemical agents in case they were needed.

WORLD WAR II

The varying landscapes and theaters in which the Second World War would be fought convinced military planners early on of the need for an effective medical strategy. Practically every disease known to modern medicine would be confronted by the Allied soldiers, including bubonic plague, malaria, dengue fever, and sandfly fever. Patriotically, the Rockefeller Foundation offered to vaccinate the majority of US servicemen against yellow fever at no charge to the government. From January 1941 to April 1942, seven million GIs were inoculated with the 17D version of the vaccination. Unfortunately, some of those immunized received a tainted batch. Around 27,000 soldiers came down with Hepatitis B, including Gen. Joseph Stilwell, who suffered from the virus during his hellish trek through Burma, making his accomplishment that much more noteworthy. The *Chicago Tribune* was one of the few papers that actively sought to investigate the matter, complaining that the victims of the mistake were twenty times the number of battle casualties up to that point in the war. Yet the rest of the nation's media and citizens seemed willing to downplay the issue as an example of wartime necessity. Initially the majority of victims thought that their sudden yellowish tinge was from the weakened virus within the vaccine, only to discover later the true cause.[2] Overall though, despite these few occurrences the program was a complete success, with the American military suffering no reported cases of the illness during the war. By 1942 all military personnel were also receiving vaccinations for typhoid, smallpox, and tetanus as well. The government was committed to removing the biggest killers of troops from the battlefield.

Both the Germans and Americans attempted to research and rush through vaccinations and treatments for the various diseases that would arise during the war. For the Nazis these involved the tests of Claus Schilling who experimented with malaria medicines and typhus treatments on prisoners at Dachau. Similar trials took place at Buchenwald and among Russian prisoners of war. Due to the over one thousand deaths that resulted from his work, Schilling was executed by the Allies after the war. The

American government ran similar operations among the mentally ill, infecting them with malaria to provide blood samples. Soldiers in the field were then used to test experimental treatments. The Allies eventually acquired German anti-malarial drugs, originally developed by Bayer in the 1930s, which were mass-produced as chloroquine. These were immoral experiments which cost thousands of lives but which saved perhaps millions.

Overall the nation's handling of illness during the various campaigns of the war was largely successful. Though the war lasted much longer than World War I and involved almost four times as many soldiers, deaths from disease were miniscule. For example, whereas there were 102,000 cases of measles in World War I with 2,370 deaths, there were only 60,809 cases in World War II with only 33 deaths reported. In addition, 90% of the cases in the latter war erupted back home in the United States and not in the field. In a similar fashion while there were 135,830 incidences of rubella, over 125,000 of these occurred in training camps in America.[3] Likewise the military suffered less than half of the cases of mumps that it did during the Great War, and only a fraction of the instances of pneumonia. The famed "transport cold" remained the single widest-spread contagion to afflict troops transiting to the battlefield. Cramped conditions, poor ventilation, and new microbes led to high rates of simple upper respiratory infections. Some reports suggest that as many as 80% of all raw recruits contracted a cold or other simple virus while sailing across the Atlantic Ocean.[4] On a positive note, however, the "transport cold" did help to season troops against England's infamous weather, which would have undoubtedly led to additional illnesses upon prolonged stationing in that country.

The two areas of the war that saw the greatest epidemiological activity were the North African-Mediterranean and South Pacific theaters. Though military planners had assumed that the landings in North Africa during Operation Torch were to be largely unopposed by troops, they were conscience of the determined local resistance to be offered by disease. Malaria, respiratory and intestinal diseases, as well as venereal diseases, were all known to be rampant in the region. Thus the American, British, and French forces undertook concentrated efforts to reduce the possibility of an epidemic outbreak. Field Orders issued for the 1st Infantry Division mandated, "a special inspection twenty-four (24) hours prior to debarkation, special emphasis to be paid to skin diseases, parasitic diseases, contagious, and infectious diseases."[5] Thanks in large parts to these efforts, disease bore little impact upon the Allied campaign to defeat the Afrika Korps.

On the other hand, the Italian and German soldiers who had been sent to North Africa did suffer for years due to the diseases of the region. The former Italian Chief of Staff, Mario Berti, who was nominally in charge of the 10th Army in Libya before the arrival of Rommel's Afrika Korps, was frequently taken ill. His most inopportune sickness came in December of 1940 when the British launched Operation Compass and pushed the Italians from western Egypt. More devastating for the war plans of the Axis, though, were the health issues of Erwin Rommel. Following his failure at the Battle of Alam Halfa in early September, Rommel had departed for medical care in Germany. A month into his convalescence, General Bernard Montgomery launched a massive offensovee against the Germans and Italians which became the pivotal Battle of El Alamein. General Georg Stumme, who had replaced Rommel, failed to hold back the assault and then died of a heart attack on October 24, 1942, only a day after the beginning of the battle. Rommel had to be rushed back to Africa, arriving there on the 25th. Thanks in part to Stumme's failure to commit his artillery, his apparent death from cardiac arrest, and the gap in leadership caused by the delayed return of Rommel, the German and Italian forces in North Africa began to fall back. The Battle of El Alamein served as the turning point of the war in North Africa, starting the general Axis withdrawal which would only quicken with the landing of American forces in the west. Heart disease had helped to liberate North Africa.

Similar fears erupted in the transit from Tunisia to Sicily in 1943. Sandfly fever, an arthropod borne phlebovirus, was known to be endemic to the region. A highly contagious illness, some military hospitals in the Middle East reported infection rates of 100%, while 80% of the headquarters staff at Caserta came down with the virus as well.[6] Over 20% of German troops in Athens had in fact been rendered combat ineffective between July and August of 1941 due to an outbreak of the virus. Likewise, Rommel's offensive into Egypt in 1942 was aided by an outbreak of the illness thanks to the blast walls constructed around many of the British tents, as these supplied perfect breeding niches for the sandflies.[7] Though the disease did reach epidemic proportions on Sicily by August, it did not have a serious impact on the campaign.[8] Proper education of the soldiers involved in the invasion, and judicious use of DDT helped to keep infection rates at a low level. The disease that did prove to be a danger to the Allies, though, was malaria. During the conquest of the island, American and British forces lost the equivalent of two whole divisions to the contagion. In total over 21,000 soldiers were stricken with the illness, while during the same time period the two armies suffered only 17,000 wounded in battle. In fact Charles H. Kuhl, the soldier slapped by General Patton on August 3, 1943, wasn't actually suffering from fatigue or shellshock at the time, but malaria.[9] Overall there were 69,000 cases of malaria reported in North Africa, Sicily, and southern Italy from 1943 to 1944, resulting in the loss of both manpower and time.

The transit from Sicily to mainland Italy brought similar problems and concerns. Besides the fear of malaria, which continued to confront the Allies, typhus had become a larger strategic issue. In 1943 a massive epidemic of the illness emerged in German-occupied Naples. The disease was brought by infected soldiers who had been transported by train from the Russian front to Italy in February, leaving over 500 of their number hospitalized in both Foggia and Bari along the way. By March 1st incidents of typhus were reported in Naples. The contagion quickly spread through the slums, the prisons, and thanks to the overcrowded bomb shelters, the general populace as well. Allied bombings and the forced evacuation of 300,000 citizens and prisoners by the Germans in late September only further spread the disease. The Allies were thus confronted with a dangerous situation when the first British soldiers rode into the city on October 1, 1943, with over 2,000 cases reported by the spring of 1944.

Fortunately for American and British war planners, the USDA had developed and tested MYL, a pyrethrum containing lousing powder, in Florida back in 1942. The next year, the Typhus Commission set up by the Surgeon General demonstrated the effectiveness of the powder in trials carried out in Egypt. The Rockefeller Foundation's health commission then initiated the critically important program of delousing Naples in December of 1943. Spraying 400 lbs. of DDT by air and delivering 500,000 2 oz. tins of MYL, the United States was largely able to bring an end to the epidemic by May of 1944, having suffered only negligible casualties.[10]

Though the German tactic of releasing infected prisoners and driving out the population can be considered biological warfare, a far more obvious version was launched in January of 1944. With the Allied advance on Rome slowed by the German Gustav Line, Prime Minister Churchill, himself recovering from pneumonia, had devised a daring amphibious attack around the flank of the Germans to be followed by a lightning assault on Rome. The region was dominated by the Pontine Marshes, a malarial filled swamp that had protected Rome from invasion for two thousand years more effectively than the walls of the city. Though Mussolini had famously cleared the region back in the 1930s, the German defenders sabotaged the waterworks, re-flooding and re-infecting the region. Mosquitos were even imported in early 1944 in an attempt to increase the spread of the disease. Cases of the disease in the local population rose from 1,217 in 1943 to 54,929 in 1944. Yet thanks to the liberal use of 500 gallons of DDT supplied by Merck, the Allies were able to avert a catastrophe.[11]

The invasion of Normandy and the subsequent push through Europe was likewise threatened by disease from its onset. Tuberculosis outbreaks had occurred in England since 1931 in odd numbered years, with a much larger wave assaulting the island every four years. Continuing forward from 1931 this would mean that England was due for a major tuberculosis epidemic in January of 1945. This would have undoubtedly spread along the supply routes to the Allied soldiers at the front approaching the Rhine River. In addition the already strained hospitals and food supplies in England would have been pushed beyond the breaking point had the epidemic occurred. In another stroke of luck for the Allies, the scheduled January 1943 outbreak was delayed by almost 10 months, pushing back the 1945 epidemic as well. The Allies were saved from fighting a possibly more brutal foe than the Germans. Over the course of the war 6,500 G.I.s would die from tuberculosis, with another 16,000 discharged, but not the tens of thousands that could have possibly fallen victim.

In a similar fashion, reports began to circulate that the invasion of Normandy would be confronted not just by pillboxes and barbed wire, but the defensive use by the Germans of botulinum toxin as well. To confront this threat the US government ordered the production of over one million doses of botulinum toxin vaccine. Surgeon General Parran argued for the immediate vaccination of all invading Allied soldiers. Yet the theater medical staff, consulting with General Eisenhower, decided against the idea, holding the vaccines in reserve. The rumor of the Germans using biological weapons proved, luckily for the Allies, to be false.[12] In order to be prepared, however, the landing force at Normandy included 8,000 doctors, 600,000 doses of penicillin, 50 tons of sulpha drugs, and 8 tons of medical equipment.[13]

Yet that is not to say that the Germans didn't use biological weapons. Biologist Erich Traub, who was later brought to the US after the war as part of Operation Paperclip with fellow Nazi bio-warfare scientist Kurt Blome, detailed some of the extents to which the Germans allegedly went to win the war. Under the guise of a cancer research program, Traub helped the Germans to develop numerous biological weapons. Reminiscent of actions during World War I, the Luftwaffe sprayed weaponized hoof and mouth disease over cattle in Russia, though with little success. According to famed Soviet defector Ken Alibek, the Russians countered these German actions by utilizing their own biological weapons. During the pivotal Battle of Stalingrad, Alibek alleges, the Soviets released tularemia, a bacterial infection, to cripple the German forces around the city. Though the disease did impact an estimated 75% of the population in and around Stalingrad, most sources doubt Alibek's contention of a controlled release by the Russians, in favor instead of a natural outbreak.[14] Regardless of its origins, the disease weakened the German forces in the region. At the same time an outbreak of Q fever struck Nazi soldiers in the Crimea, and typhoid caused as many casualties as battle wounds both on the Eastern and Balkan fronts for the Germans. Pestilence was in part to blame for the halting of the German advance to the East much as it had been for Napoleon's army a century and a half before.[15]

America suffered its largest onslaught of infection in the Pacific Theater. Disease bred naturally in this tropical region, and the two biggest killers that emerged were dengue fever and malaria. The US Army suffered over 84,000 cases of dengue fever, with over half of these occurring in the southwest Pacific in 1944 and 1945 alone. From March to May of 1942 an epidemic in Queensland and the Northern Territories affected 80% of the troops in the region. The disease then followed the Allies as they worked their way through the island chains up towards Japan. In February of 1943, 25% of the Allied troops on the New Hebrides and New Caledonia had fallen victim to dengue. Much of the outbreak was caused by the soldiers' own lack of sanitation, creating a breeding ground for mosquitos by the

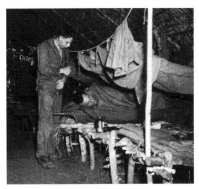

Malaria hospital in New Guinea, 1943.

improper disposal of old containers and tins. By 1944, 24,000 cases had emerged in New Guinea, with another 11,000 reported in 1945 in the Philippines. Likewise, barely a month after the landings on Saipan in July of 1944, dengue fever erupted with a fury among the Americans on the island. Over 20,000 cases were reported, nearly a third of the invading force, threatening to imperil the seizure of the island.[16] As well as around encampments, the insects were able to breed in the various shell holes and rubble that littered Saipan's beaches and jungles. Yet thanks to the spraying of over 8,600 gallons of DDT in September, the disease was finally brought under control. Troops on the island were split into distinctive areas, the commanding officer of each being responsible for the eradication of mosquito breeding grounds in the area.[17]

The US military has had operational familiarity with malaria stretching back to the American Revolution. Yet the extended time that the army and navy would stay in the tropical Pacific region worried many medical men and military planners. Contact with the disease started early for the Allies. During the Japanese invasion of the Philippines it is estimated that 24,000 of 76,000 American and Filipino soldiers were stricken with malaria. This proved to be a boon to the Japanese invasion force as it was only around 57,000 strong. The commanding officer of General Hospital No. 2 wrote to his superiors that . . .

> If the malaria situation is not brought under control, the efficiency of the whole Army will be greatly impaired; in fact it will be unable to perform its combat functions. It is my candid and conservative opinion that if we do not secure a sufficient supply of quinine for our troops from front to rear that all other supplies we may get, with the exception of rations, will be of little or no value.[18]

As American forces withdrew to the Bataan Peninsula, a region rife with the disease, conditions worsened. By March of 1942, quinine had run out for the defenders of the Bataan Peninsula, forcing a local commander to order that no Marine be excused from duty unless he possessed a fever above 103°.

In the little-recognized campaign to retake the Aleutian Islands from the Japanese, fought from 1942 to 1943, American troops battled perhaps more familiar diseases, but nonetheless suffered extensive casualties. Cold temperatures and the outbreak of disease severely impacted the attacks against Attu and Kiska. During the campaign to take the former island, the United States would lose 580

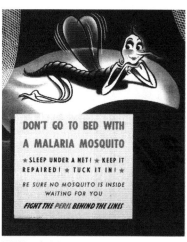

WWII malaria poster.

men from enemy action and 614 from disease. Likewise, though the Japanese had abandoned Kiska Island, 313 Americans died resecuring it, mostly due to disease and frostbite.

The Allied offensive to regain the islands of the South Pacific was likewise impacted by disease. This was especially true during the campaigns on Guadalcanal and in New Guinea. From March to December of 1943 alone, the U.S. saw over 10,000 cases of malaria on Guadalcanal. When over 80% of the 1st Marine Division was struck down, the entire unit was declared no longer combat effective. This stands in stark contrast to the 4% of the division that died taking the island. Similar stories unfolded with other units on the island as well. The Americal Division had to be removed from Guadalcanal and sent to Fiji for recovery after only five months on the island due to losses from malaria. The 25th Division would suffer a similar fate on Guadalcanal, losing 46% of its strength by December of 1943.

New Guinea also proved to be a biologically dangerous battlefield for the Allies. The U.S. 32nd Division lost 67% of its men to malaria as opposed to only 26% of its strength to wounds or death. During the campaign on the island American units suffered a ratio of 5:1 casualties from disease to battle wounds. Overall it is estimated that between 60–65% of all Americans in the South Pacific contracted malaria at some point during their time in the military. From 1942 to 1945 this amounted to over 500,000 cases.

Yet the Japanese were not immune to the tropical diseases that beset

US Army Sign in the Pacific Theater, 1941.

the Allies either. In fact, as the Japanese occupied the islands for longer periods of time than the Americans, they tended to suffer more. Some units reported ineffective rates of up to 90%. This was especially true on Guadal-canal, where disease crippled the offensive plans of General Hyakutake and others. The Japanese were unable to hold off the American invasion or retake the islands due to the illnesses that crippled their forces. Around the same time the Japanese thrust into Burma was ultimately defeated and turned back mainly because of disease. In New Guinea, while the Americans were suffering from malaria, so were the Japanese. In fact by January of 1943, only 1 out of 7 Japanese soldiers was fit for duty.

In fact, the American victory in the Pacific owed as much to its handling of disease as its handling of aircraft carriers. Repeated and effective use of DDT allowed for the cleansing of dozens of island chains, depriving the protists that cause the disease available hosts. As well, though American supplies of quinine began to dry up after the conquest of Java by the Japanese, the government pushed through the production of the anti-malarial drug Atabrine. Though it tended to turn the skin of GIs yellow and caused diarrhea and allegedly sterility, the drug proved to be very effective in the fight against the disease.[19] The military went to great lengths to encourage soldiers to take the disease seriously. Among those employed by the government to

WWII health poster.

aid in anti-malarial efforts was Theodor Geisel, better known as Dr. Seuss, who drew cartoons about "Ann," the Anopheles mosquito, during the war. Additionally, soldiers were subjected to films, pinup calendars, and the popular cartoon "Malaria Moe." Later stories assert that DDT became so popular among the troops that it was mixed with gin to create a Mickey Slim, a drink that would, "give you a feeling of happiness and merriment."[20] In the end, America's chemical expertise was as responsible for its victory in the Pacific as its technical or nuclear advancements.

Wyeth and Brothers poster, 1942.

Finally, venereal disease remained a common threat to all militaries in the war. In June of 1940 the American government formed a subcommittee on venereal disease, with the Surgeon General issuing a variety of circulars detailing containment and treatment. Thanks in large parts to the presence of penicillin during the war, outbreaks became largely treatable. Still, over 200,000 cases of syphilis alone were reported between 1941 and 1945 by Americans in uniform. Costing both money for treatment and lost man-hours, VD, though not deadly, did hinder the war effort. Much as during

DDT shipment to Iran, 1944.

World War I, rapid-treatment centers were set up and local prostitutes detained in order to combat the spread of infection. Thanks to better treatment, public education, a moral crusade against prostitution, and preventive measures, including the free dispersal of prophylactics to soldiers, the United States was able to cut its incapacitated rate 30-fold from World War I.

In terms of biological warfare, the Americans utilized the tactic little more than they had in the previous war. The Geneva Protocol of 1925 had outlawed the use of such weapons as part of the world's general disarmament movement of the 1920s, and most belligerents respected the treaty. Yet there was palpable fear in the administration that either Japan or Germany could employ biological warfare against the United States. J. Edgar Hoover in fact warned the President in 1942 that German scientists were working closely with their Japanese counterparts for just such a campaign. In this he was undoubtedly aware of the German terror campaign of 1914–18. Secretary of War Stimson had already tasked the Army's Chemical Warfare Services with starting the research and production of biological agents and counteragents in 1941, when in 1942 he broached the subject with Roosevelt. The President agreed with Stimson's recommendations, and that same year the War Reserve Service was established under George Merck with a budget of $25,000,000. Camp Detrick in Frederick, Maryland would become the headquarters of the nation's biological weapons program, with all testing eventually carried out at Dugway Proving Grounds in Utah. While most of the war was spent investigating the weaponization of anthrax and botulism, with the British and American armies requesting over a million anthrax bombs, other diseases were experimented with as well. These included brucellosis, psittacosis, tularemia, and glanders. In addition scientists looked for ways to cripple Japan's agricultural production in an effort to starve them into submission. By 1945 Detrick held more than 245 structures and employed a staff of over 2,200 military personnel and civilians. Yet Roosevelt never authorized the use of any of these weapons in anger against either the Germans or Japanese; it appeared that the days of smallpox blankets were over.

In fact the only nation during the war to actively use biological weapons on a large scale was Japan. The infamous Unit 731 had tested a variety of biological agents upon American and Chinese prisoners of war for years in various camps located in Manchuria. With these weapons in tow, at least

11 cities in China became targets for biological attacks during their offensive. The Japanese released bubonic plague, typhoid, and cholera directly into civilian populations in an effort to destabilize the country. On October 4, 1940, plague-infected fleas were dropped over Quzhou, Zhejiang, resulting in over 2,000 deaths over the course of the next year. It has been estimated that over 200,000 Chinese civilians would perish during World War II from biological warfare alone. In one of the most far-fetched schemes or rumors of the war, it was alleged that the Japanese were planning to deliver biological agents to American soil by balloons. Over 9,000 balloon bombs had in fact been released by Tokyo during the war, though few reached American soil, and far fewer still caused any damage. Yet the fear of the possibility remained in the minds of American planners. The Japanese themselves produced 400kg of anthrax by 1945 for use in bombs, though ultimately none was ever used. Finally in 1945 the US Navy captured an enlarged Japanese I-400 submarine that had been modified to carry Seiran planes loaded with biological weapons against the West Coast of America. Again, Japan's appreciation of biological warfare served as a wakeup call to postwar American planners.

Overall, World War II represented the turning point in the military's war on disease. For the first time in American history, conflict fatalities resulted more from battle deaths than from illness. Two-thirds of all deaths in the war were non-biological in origin, a vast departure from the Civil War. Only the dedicated efforts of the military, its medical branches, and the increased sanitation efforts of the soldiers themselves helped to bring this about. The United States had taken on and defeated not only fascism but pestilence as well. Still, with over 100,000 recorded deaths from pestilence, it was a battle that was being won, but that had not quite yet come to a conclusion.

THE COLD WAR

One of the most damaging illnesses of World War II was the heart disease and possible cancer that was affecting Franklin Delano Roosevelt. The President's health had never been good, and a massive cover-up had accompanied him during his 12 years in office. In fact privately his doctors argued against his running for a third term in 1940, let alone a fourth in 1944.[21] In February of 1945 he attended his last wartime conference with Joseph

Stalin, in which he sought to convince the Soviets to attack Japan once Germany had surrendered. He was visibly ill at the time, with Ambassador to China Patrick Hurley later recounting that Roosevelt was, "a very sick man at Yalta."[22] Though debate still rages as to the extent that his condition played in the talks, after only a thirty-minute meeting with Stalin he agreed to turn over Mongolia and much of Manchuria. Likewise Stalin was largely given a free hand in the rebuilding of, and establishment of governments in, Eastern Europe. These land grabs by the Soviets helped to set the stage for the half-century Cold War that was to follow.

The conflict that raged between the United States and the Soviet Union was fought by science as much as by political philosophy. An arms race quickly developed between the West and East not confined to only nuclear weapons, tanks, and bombers. A quest to develop both biological weapons and countermeasures would consume both countries for the next 25 years, and push the United States government to new heights of political and industrial power as well as new ethical lows.

In a similar way to the country's rocket and space program, America combed through German and Japanese records and scientists to help propel forward its biological weapons program. As part of Operation Paperclip in Europe, dozens of Nazi scientists, engineers, and spies were granted amnesty and a new life in the U.S. to keep them out of the hands of the Soviets. The most famous Japanese researcher picked up in this process was the bioweapon specialist Shiro Ishii. A former member of the notorious Unit 731, Ishii was granted amnesty by Washington in 1947 to not only acquire his knowledge but also to keep him from being tried with other members of his former unit by the Soviets at Khabarovsk. Though the information garnered from him was later proven to be of little value, Ishii continued to work with the American government up until his death in the late 1950s.[23]

Fort. Detrick would continue the work that it had started during the Second World War to construct effective biological weapons. Some of the bacteria and viruses weaponized or researched included tularemia, smallpox, anthrax, botulism, brucellosis, Q fever, Hantavirus, ricin, yellow fever, bubonic plague, potato blight, ricin, Bolivian hemorrhagic fever, Argentine hemorrhagic fever, and various anti-crop and anti-animal contagions. The impetus for this program arose in part from the Baldwin Report of 1948. Congress had formed a committee under Dr. Ira Baldwin, one of the top scientists at Fort Detrick Maryland, to investigate biological weapons and warfare. The report detailed that American transportation, ventilation, and water supply systems were shockingly susceptible to biological warfare. Testing needed to be done both to gauge flaws and prepare researchers and the government to react to such an outbreak. Ideally, these tests needed to be done in secret to fully realize the effects of the introduction of a biological contagion into a major city in the nation. Thus began a decades long covert and controversial program by the government of the United States to test biological agents, both harmless and deadly, among its citizens. All of this hinged on the knowledge that the Soviets were likewise developing these weapons and would potentially use them to covertly or overtly attack the West.

The Special Operations Division of the Army's Chemical Corps at Fort Detrick introduced what at the time was thought to be a harmless bacterial agent, *Serratia marcescens*, into the ventilation system of the Pentagon in August of 1949 to test how quickly and effectively it would spread. The next year, a harmless anthrax stimulant was sprayed on Norfolk, Hampton Roads, and Newport News by the US Navy. A similar operation took place off the coast of San Francisco in which a three-mile-wide cloud of *Serratia marcescens* was released towards the city for six straight days. Scientists estimated that the vast majority of the city's population breathed in the bacteria. In all from 1949 to 1969 it is surmised that well over 240 such open air tests were performed across the nation. As mentioned above, though *Serratia* was thought at the time to be harmless, in actuality it can lead to several types of infections. Claims have since been made that 11 residents of San Francisco became ill following Operation Sea Spray and that one of these patients, Edward Nevin, subsequently died of a rare urinary tract infection.[24]

On August 9, 1952, a mock city constructed at the Dugway testing facility in Utah and inhabited by over 2,000 guinea pigs was attacked using brucellosis. Considered a success, the United States government expanded its testing program the next year to include actual American cities. In 1953 the St. Jo Program launched a mock anthrax assault on St. Louis, Minneapolis, and Winnipeg, dispensing stimulants from vehicle mounted sprayers. During that same year the Air Force officially adopted brucellosis as its standard biological weapon, and a secondary production facility to produce cluster bombs filled with the illness was constructed at Pine Bluff, Arkansas.

Yet, as bombs were not always the most effective way to transmit diseases, the government also hoped to utilize insects to deliver contagions, as they have done so naturally and effectively for thousands of years. Thus a variety of tests were carried out to determine the feasibility of employing entomological weapons. Operations Big Buzz (1955), Drop Kick (1956), and May Day (1956) saw the bombing of rural Georgia using E14 cardboard munitions containing hundreds of thousands of mosquitos. Operation Big Itch, launched in 1954, used a similar method to drop hundreds of thousands of fleas in Utah. Overall the projects were immensely successful and demonstrated a novel way to introduce a variety of contagions against an enemy in war.

One of the most controversial tests performed by the U.S. was Operation Whitecoat, which lasted from 1954 to 1973. During this almost two-decade period, some 2,200 persons were exposed to a variety of viral and bacterial toxins. The test subjects were for the most part Seventh Day Adventists, who though drafted for duty refused to serve in combat. The Army instead offered them the option to serve as subjects in biological tests. Refraining from alcohol, nicotine, and caffeine, these men were in ideal health and would provide unblemished test data. Placed in the "Eight Ball," a gigantic sphere developed at Fort Detrick, the men would be fitted with masks and exposed to such contagions as tularemia or Q fever. On July 21, 1955 alone, 30 such men contracted Q fever, though all were given antibiotics and made full recoveries. In many ways, though controversial, Whitecoat was actually quite ethical in its use of informed consent from its subjects.[25]

The government launched Operation Large Area Coverage in December of 1957. The aim of the project was to test the extent to which airborne particles would travel across the country. To accommodate this research, airplanes sprayed zinc cadmium sulfide as they flew on predetermined flight patterns. The first such experiment resulted in a zone of infection stretching from South Dakota to Minnesota, with some particles being detected over 1,200 miles away. Later tests were conducted spraying test particles from Ohio to Texas and Michigan to Kansas, covering in total over 33 urban and rural areas. These experiments would continue into the beginning of the 1960s.

Research and human experimentation picked up during Kennedy's and Johnson's presidencies. One of the largest such programs was Project 112, a four nation operation involving the increased testing of chemical and bio-logical agents as an alternative to nuclear weapons. Various contagions, defoliants, and irritants were tested in such diverse environments as Alaska, Panama, and Okinawa. During Operations Devil Hole I and II, sarin and VX gas were released in Alaskan forests near Fort Greeley to test their range and effectiveness. In 1965 the U.S. released tularemia in a test on Johnston Atoll in the Pacific, and in 1969 spread bacterial agents over a 900-mile swath of Einewetok Atoll. Simulated attacks also continued to be carried out in various cities around the country. In May of 1965 one such "assault" took place on the bus terminal and airport in Washington D.C. Likewise, in June of 1966, bacteria-filled light bulbs were used to test spread patterns in the New York subway system. During Operation Big Tom, *Bacillus glo-bigii*, a relative of anthrax, was sprayed over the island of Oahu. Similar tests were also carried out in Canada and at various points in the UK. In the closely linked Project SHAD (Shipboard Hazard and Defense), the Penta-gon sprayed chemical and biological agents on over 40 ships during the 1960s. Some of these experiments also tested the offensive use of mosquitos at sea.

Information about the true scope of the nation's storage and testing of biological agents came to light in 1969 following the accidental release of VX gas at a facility in Okinawa. Outrage by locals and the Japanese gov-ernment led the United States in Operation Red Hat to remove all of the chemical and biological weapons stored there, totaling 12,000 tons, to John-ston Atoll.[26] More recent evidence revealed that American soldiers had been exposed to numerous defoliants and chemical agents for years upon the island, possibly leading to long-term health issues and even death. A similar disaster unfolded at Dugway in 1968 when VX was accidently released, leading to the death of 6,400 sheep.

Plum Island, off the eastern coast of Long Island, also served as a bio-logical testing facility for the government. Various tests were conducted at the highly secret location connected to animal related illnesses, including hoof and mouth disease and anthrax. The installation holds the dubious honor of being accused by various researchers and authors of being respon-sible for the spread of numerous diseases to the mainland. A spike in Lyme Disease cases after 1975 in New England and the arrival of West Nile Virus in America in 1999 have all been tentatively blamed upon accidents at the facility. Though access to the area is restricted, Plum Island also serves as a bird sanctuary, leading some to question whether they could have served as vectors for a variety of contagions not normally associated with the region.

In terms of actual usage in combat, the American experience tends to be based more on rumor and propaganda than actual truth. Both North Korea and China alleged that the U.S. utilized biological weapons during the Korean War. Outbreaks in the winter and spring of 1952 of bubonic plague, cholera, meningitis, and smallpox in North Korea were blamed on the spread of these contagions by America. The Communists alleged that Washington used beetles and other insects to introduce the plagues into the Korean peninsula. Similar claims followed an outbreak of smallpox in Pyongyang and a bacterial outbreak in central China. In response, China even established the International Scientific Commission, heavily staffed with fellow travelers and brainwashed POWs, to investigate its claims. Chairman Mao himself claimed that the United States, *"used bacteriological weapons eight times, from planes and through artillery shells."*[27] Though it has become apparent from documents obtained from the Kremlin after the fall of the USSR that this was simply propaganda on the part of the Communists, it was still treated as genuine by the Chinese at the time, and used as an impetus by Mao to improve the health and sanitation of the country through the "Patriotic Health and Epidemic Prevention Campaign." Rumors would again circulate in the 1960s around the use of biological agents in Cuba by the Kennedy administration in an attempt to destabilize Castro's regime. Possibly as part of Operation Mongoose, this allegedly involved the poisoning and destruction of sugar crops and livestock to weaken the island's economy.

The entire biological warfare program of the nation came to an abrupt end in 1969. After a promising start, two decades of research and tests, and after the NSC had rescinded its anti-first strike policy in terms of biological weapons, President Nixon ended the program. By 1969 the bioweapons division of the military was costing the taxpayer around $300 million annually. This vast expenditure at a time of a stagnating economy, combined with Nixon's moves to lessen tension in the Cold War, and no doubt in part due to the untested benefits of the program, led the president to renounce biological warfare. In 1972 a biological weapons convention was signed by over 79 nations, with the US Senate approving of it in 1974. Five years later the Belmont Report was published, summarizing the guidelines for using human subjects in any future experimentation, promising an end to decades of surreptitious and unethical procedures.

Though the United States may have abandoned its biological weapons program, its enemies did not. An anthrax outbreak in Sverdlovsk in 1979 that killed at least 68 was suspected at the time as due to an accident at a nearby biological weapons facility. Though vehemently denied by the Soviets at the time, President Boris Yeltsin later admitted to the fact in 1992.[28] Over 50,000 residents were vaccinated, buildings were scrubbed down, and roads were repaved. In another incident, the American government accused the USSR in 1981 of supplying T-2 mycotoxin to Communist states in Southeast Asia, which had been used to possibly kill thousands of refugees during the 1970s. The accusation was heavily disputed then and since. An international team sent to the region ruled the "yellow rain" gathered as evidence was most likely honeybee feces. Yet the American government stood by its claims of the use of biological weapons by the Communists.[29]

During that same decade in an undisputed incident, Iraq employed chemical and biological weapons to kill tens of thousands of Iraqis, Kurds, and Iranians. To protect its military personnel from both naturally occurring diseases as well as potential biological threats, the United States government mandated a number of vaccinations be given prior to the launching of Operation Desert Storm. Among these were inoculations against yellow fever, typhoid, cholera, hepatitis B, meningitis, whooping cough, polio, and tetanus, which were received by all Americans sent to the region. An estimated 150,000 men also received vaccinations against anthrax and another 8,000 were protected from botulism toxin. Following the Persian Gulf War, the Veterans Administration estimated that 25–32% of soldiers who had served in Iraq and later operations suffered from Gulf War Syndrome. This physical and psychological disease remains of unknown origin, with some suggesting that Pyridostigmine Bromide given to soldiers to combat chemical weapons exposure or some other source may have been the cause.[30] In fact the continued presence of these chemical and biological weapons in Saddam Hussein's arsenal would serve as the strongest argument given by the George W. Bush administration for the second war against Iraq a decade later.

The various military campaigns of the Cold War saw a dramatic decline in the impact of disease upon the military. Casualties for the Korean War, while totaling over 54,000 men, saw only 2,835 deaths from "other" causes, including accidents and disease. The new Mobile Army Surgical Hospitals played a distinctive role in helping to reduce bacterial infection rates among those admitted with battle wounds. As for Vietnam, besides instances of intestinal illness and malaria, skin diseases due to tropical conditions proved to be the most serious issue for medical staff. And though deaths from disease remained in a steady decline since World War I, man-hours lost for those admitted to military hospitals with curable diseases tripled those lost to battle injuries in 1965 before leveling out by 1970. Of the 16.5 million days of active duty lost between 1965 and 1970, 8.1 million were due to disease.[31] Extrapolated, this amounts to 3,742 men incapacitated per day due to disease alone.

Venereal disease though, made a startling resurgence beginning with the Korean War. Perhaps due to a combination of the sedentary nature of much of the conflict along with American soldiers fighting on friendly soil, frequent fraternization with local women led to a reported rate of infection at three times the level of that of World War II. This trend line increased in Vietnam where infections were now six times that same level.[32] Perhaps the sexual revolution being experienced back home pushed up these occurrences as well. By the mid-1960s the incident rate for venereal infections at army bases in the United States was at 325 per 1,000 men.[33] As always, the spread of these diseases to the general population once the soldiers returned home potentially helped lead to the spike of such diseases over the next two decades.

The Cold War pushed the United States to the status of a global superpower. The nation was tested militarily, socially, economically, scientifically, and politically for almost five decades. During this time, issues of biological weapons, human testing, and national security further tested the moral and constitutional limits of the country as well. Though no major biological exchange ever did take place between East and West, the tests themselves and the stockpiles of weapons would present problems for decades to come. Likewise, though the population was never exposed to disease by force of arms, it did experience an upsurge in pestilence during this time period due to its own evolving social climate, an issue that affected the West more than the East.

CHAPTER THIRTEEN

AL-QAEDA, ANTHRAX, AND AMERICA
TERRORISM AND DISEASE IN POST-COLD WAR AMERICA

*"The use of chemical and biological weapons
against population centers is allowed and
is strongly recommended."*
—ANWAR AL-AWLAKI, 2011

With its equal potential for producing both death and fear, disease became a natural weapon for a variety of terrorist groups fighting against the United States. In many ways it was a weapon that was easier to manufacture and transport than more traditional arms and explosives; a weapon of great magnitude that could enter the nation practically undetected. For any organization which built its methods around the concept of propaganda of the deed, the fear and destruction that could be wrought by the release of a biological or chemical agent was invaluable. It was in part the nation's own conquest of pestilence that made this such an effective weapon. The importation of disease in the 19th and early 20th centuries, besides providing fertile ground for Nativism, did not generate the same level of outright terror as it would after the ending of the Cold War.[1] Fear is directly relatable to both the strength and rarity of an action, and in an age where even chicken pox is moving towards extinction, biological horrors become nightmares.

Beginning in the 1970s, rumors of biological attacks began to manifest in America. One notable example was a story of the Weathermen Underground attempting to procure chemical and biological weapons from Fort Detrick. Though this allegation was never proven, two other terrorists, Steven Pera and Allen Schwander, were reported by authorities in 1972 to have within their possession *salmonella typhi*. Belonging to an, until then, unknown group named R.I.S.E., Pera and Schwander spoke of eliminating humanity through disease and repopulating the world, possibly due to ecological concerns.[2] The two eventually fled to Cuba and their "planned attack" was never substantiated. One of the first confirmed incidents of bioterrorism in recent history involved employing salmonella in a concentrated effort to seize political power. Bhagwan Shree Rajneesh had relocated to Oregon from India in the 1980s and began to build up a cult following around his controversial New Age ideas. Effectively seizing control of Antelope, Oregon, the group renamed it Rajneeshpuram. However, many in the county became concerned with the growing power of the cult and sought at every opportunity to restrict its growth. Members of the cult settled upon a plan to win elections in Wasco County in 1984 by poisoning the electorate, thus restricting voter turnout. Salmonella was secreted into the water of two county commissioners and introduced into a series of salad bars at various restaurants in the Dalles. In the end 751 citizens would become sick, with 45 requiring hospitalization. Yet the attack was a disaster. Suspecting the sect of being responsible from the beginning, voter turnout actually increased, resecuring control of the county for the local population. Though the cult did not achieve its goal, the incident stands as the first large-scale biological terrorist attack on American soil.

Only a week after the terrorist attacks of September 11, 2001, a series of letters containing deadly anthrax began to show up in various news agencies and in the offices of two US Congressmen. The letters, sent on Sept. 18th and Oct. 9th, would end up killing five persons and seriously sickening seventeen. Panic gripped the nation as a suspect was not quickly identified and fears of connections to al-Qaeda or Iraq were suspected. Osama bin Laden himself stated in an October of 2001 interview with Al Jazeera that the anthrax attacks were, "*a punishment from God and a response to oppressed mothers' prayers in Lebanon, Iraq, Palestine, and everywhere.*"[3] Though the apparent targets, Senators Daschle and Leahy, escaped harm, security was tightened at government and media offices as well as within mail delivery centers. The strain of anthrax was eventually identified as the Ames strain, which was first isolated at Fort Detrick. A deceased scientist at the lab, Bruce Edward Ivins, was eventually declared the sole suspect after his suicide, yet many remained unclear as to the extent of his guilt.

All of this followed a news article written on September 4th detailing that the government had for years been running various biological warfare operations. Project Bacchus involved federal agents setting up a biological weapons lab from scratch to test the feasibility of terrorists doing the same. Project Jefferson had the government producing a highly deadly strain of anthrax in an attempt to test modern vaccines. Finally, Project Clear Vision involved the government testing the release capability of biological weapon bombs recently produced abroad.[4] Following the terrorist attacks of 9/11 and the anthrax letters, the Bush administration undertook Operation Dark Winter in June of 2001 to test a mock smallpox attack on Oklahoma City to gauge local and federal response.

In response to all of these incidents and tests, the federal government passed a series of acts aimed at mass-producing and stockpiling vaccines in case of a bioterror attack in the future. The first of these, the 2004 Project Bioshield Act, pledged $5 billion to produce hundreds of thousands of anti-toxins for botulinum and anthrax. The Biodefense and Pandemic Vaccine and Drug Development Act of 2005, and the Pandemic and All-Hazards Preparedness Reauthorization Act of 2013, sought to both allow for the quicker approval of vaccines by the FDA in times of emergency and to restrengthen the original piece of legislation.

The United States itself has even attempted to employ disease as a means by which to combat terrorism. It was reported in 2014 that the CIA had considered using a polio vaccine program in Pakistan in 2011 in order to gather DNA evidence of the whereabouts of Osama bin Laden. Though the operation was never carried out, subsequent Taliban and tribal mistrust of the various vaccine drives led to riots and attacks that caused the deaths of at least 56 people. The implications for world health management proved serious enough that the CIA and White House had to issue a statement in May of 2014 promising not to use disease prevention programs as a cover for covert actions in the future.[5]

Fear of biological weapons being used by either terrorists or rogue nations became one of the leading arguments for the invasion of Iraq in 2003, especially following discoveries of advanced research by Al-Qaeda in Afghanistan after the 2001 invasion.[6] Various officials in the Bush administration claimed that Abu Musab al-Zarqawi had set up a biological weapons facility at the Kermal Al Ansar camp in northern Iraq. Al-Qaeda and its affiliates have expressed interest on numerous occasions in using chemical and biological weapons to confront the United States and the West. One of the group's chief clerics, Anwar al-Awlaki, once stated that, *"the use of chemical and biological weapons against population centers is allowed and is strongly recommended."*[7] Aafia Siddiqui, who was arrested in Pakistan in 2008 for her connections to Al-Qaeda, had in her possession at the time lists of potential attack sites in America, including Plum Island, and almost two pounds of sodium cyanide. Meanwhile in January of 2009, 40 members of an Al-Qaeda cell in Algeria were reportedly killed while experimenting with the weaponization of bubonic plague.[8] Finally, an Al-Qaeda video aired by Al Jazeera in February of 2012 showed a Kuwaiti professor urging:

> Four pounds of anthrax—in a suitcase this big—carried by a fighter through tunnels from Mexico into the U.S. are guaranteed to kill 330,000 Americans within a single hour if it is properly spread in population centers there. What a horrifying idea; 9/11 will be small change in comparison. Am I right? There is no need for airplanes, conspiracies, timings and so on. One person, with the courage to carry 4 pounds of anthrax, will go to the White House lawn, and will spread this 'confetti' all over them, and then we'll do these cries of joy. It will turn into a real celebration.[9]

Khalid Sheikh Mohammad, the proclaimed mastermind of the September 11 attacks, was finally captured in Rawalpindi hiding at the home of a sympathetic Pakistani bacteriologist named Dr. Abdul-Quddis Khan. As for Osama bin Laden himself, his health issues, real or suspected, often became the subject of discussion during the 1990s and 2000s. Various media outlets and even the CIA speculated as to whether he was suffering from Marfan syndrome or a kidney disorder. Speculation as to his whereabouts was often caveated by his purported need for daily dialysis. NBC News went so far as to report that he was only months away from dying in December of 1998 due to heart disease or cancer.

The use of ricin as a weapon also escalated at this time. Unrelated attacks in both 2003 and 2013 in the United States by individuals, and in 2002 by Al-Qaeda in London attempted to use the highly toxic compound to target politicians and citizens. Though the attacks all failed they further showed the level to which the use of biological agents had devolved, in that they could be used by the individual. Average people could, with effort, employ weapons capable of killing thousands in a way unseen since the days of Cortez standing before Tenochtitlan. As the power of the individual to wreak havoc grew, so too did the size and control of government to counter the threat.

Amidst these fears of the offensive use of disease against them, the American people suddenly were confronted by the possible introduction of pestilence into the nation by their own government. As the West African Ebola outbreak raged in the late summer of 2014, President Obama approved an operational plan to send 2,800 American military personnel to Liberia on a mission to help construct treatment facilities. Operation United Assistance ran from October of 2014 to April of 2015 and represented perhaps the first time that the military had been sent abroad to combat a microscopic enemy rather than a human one. Yet many citizens were appalled by the prospect that these soldiers would become potential carriers for a hemorrhagic fever. Due to pressure from the military, President Obama finally agreed to a 21-day "controlled monitoring" of the men and women upon their return. This came in sharp contrast to his criticism only a few weeks earlier against the governors of New York and New Jersey for attempting to institute mandatory quarantines within their own states for returning nurses, doctors, and aide workers. In the end, though, the biggest enemy faced by the soldiers proved to be the rain and abysmal infrastructure of Liberia.

The rise of the Islamic State of Iraq and al-Sham (ISIS) in 2013 once again brought the issue of biological warfare by a non-state actor to the fore. This new extremist Sunni group had the advantage of attracting many young, university educated jihadists, some of whom employed their knowledge of biology and chemistry towards the production of weapons of mass destruction. A laptop captured after an ISIS force was expelled from the Syrian province of Idlib in January of 2014 contained, amongst 35,000 files, a 19-page document on the use of biological weapons. Authored by a former university student from Tunisia who had studied chemistry and physics, the paper suggested obtaining and using bubonic plague, as *"the advantage of biological weapons is that they do not cost a lot of money, while the human casualties can be huge."*[10] Also saved on the computer was a copy of a fatwa from Saudi cleric Nasir al-Fahd, which stated, *"If Muslims cannot defeat the kafir [unbe-lievers] in a different way, it is permissible to use weapons of mass destruction, even if it kills all of them and wipes them and their descendants off the face of the Earth."*[11] More recently, the Ebola outbreak in West Africa raised concerns in Spain and other nations of ISIS potentially trying to use the virus or victims of it to attack the West.

Attempts by both al-Qaeda and ISIS to obtain biological agents for use in their wars against both the West and various regimes in the Middle East and North Africa will undoubtedly continue as long as both organizations exist. Numerous studies and reports have argued that the feasibility of a bioterror attack is actually quite low.[12] Yet as history has shown, not only have more primitive forces effectively utilized disease, but the most serious outbreaks have often come about in an unintended way. Regardless of the cause of the outbreak of a particular illness, its effect can nonetheless be profound.

CONCLUSION

D isease has existed and coexisted with man since his inception, a co-habiter of the planet, a rival, and occasionally an unknowing and fickle ally. In all this time it has stood as perhaps the greatest change agent, for good or bad, for humanity. Illness has won wars, defeated armies, altered economies, produced or removed leaders, and decided the course of history. Likewise while the advance of society has seen the reduction or elimination of some diseases, it has at the same time produced or augmented others. By conquering old monsters, advancement has only created new ones.

America's experience with disease stretches back to before its founding as a nation. Disease helped to facilitate the conquest of the continent by the various European powers, and at the same time helped the English to secure their own portion of the landmass. While challenging the colonists' ability to survive, pestilence framed the mindset of the country and was perhaps the chief aid to the victory of the Americans in the Revolution. Disease followed settlers west as the nation expanded and reared its head in all major conflicts. Great steps were made by science to combat it, while at the same time the cities that housed these researchers quickly produced new illnesses. In early America, disease was an undrivable force producing both benefit and destruction, while following the Civil War, it served primarily as a catalyst for growth. The defeat of disease can be seen in a cursory examination of casualty figures stretching from the Revolution to the War in Iraq. While in the former conflict the vast majority of all deaths were from pestilence, by the time of the Iraq War disease had become largely a non-issue in the military. Despite its recent retreat, it would not be hyperbole to say that disease was the greatest and deadliest enemy ever faced by the American military. Rearing its head in almost every war, it carried off more men than did any opposing nation.

Perhaps the common theme throughout the story of civilization and disease is the greater lengths to which people will go in surrendering power over themselves to the government in exchange for protection from illness. The growth of the American government and others around the world over the past century is owed in part to the fear of the spread of contagion among its people and armies. The possibility that the nation or its economy could be laid low by invisible microbes has compelled leaders to exert more and more control over the lives of their citizens. In a world of citizen soldiers and conscription armies, the health of the population is a concern of national security. At the same time, fear among citizens of pestilence due to its now rare nature has left them more willing to accept this increasing, visible hand of government. Yet this could be a dangerous trend, with possible totalitarian results. To paraphrase Benjamin Franklin, one should never trade essential liberty for temporary security. Diseases will always be with mankind; freedom will not.

ENDNOTES

INTRODUCTION

1. Heraclitus himself succumbed to dropsy around 475 BC after his application of a liniment of cow manure failed to cure him. According to Diogenes Laertius in his *Lives and Opinions of Eminent Philosophers*, "Hermippus states, that what he asked the physicians was this, whether any one could draw off the water by depressing his intestines and when they answered that they could not, he placed himself in the sun, and ordered his servants to plaster him over with cow-dung; and being stretched out in that way, on the second day he died, and was buried in the market-place. But Neanthes, of Cyzicus says, that as he could not tear off the cow-dung, he remained there, and on account of the alteration in his appearance, he was not discovered, and so was devoured by the dogs."

2. Sun Tzu, *The Art of War*, Chapter IX, Line 12.

3. Flavius Vegetius, *De re militaria*, Book III, Section 2.

CHAPTER 1

1. Population figures for Tenochtitlan are notoriously varied, ranging from 35,000 to 500,000. For our needs we will err on the lower end of the spectrum and utilize the study done by Susan Toby Evans in *Ancient Mexico and Central America: Archaeological and Cultural History* (New York: Thames & Hudson, 2013), 549.

2. Cortes himself had upwards of 200,000 Native allies with him during the siege of Tenochtitlan.

3. A term first popularized by Julian Juderias in his 1914 work *La leyenda negra y la verdad historica*. The concept was quite popular in English history and thought from the 16th century to the present, gradually working its way into American historiography as well.

4. The smallpox outbreak was allegedly first introduced by a slave of de Narvaez when he landed to confront Cortes in April of 1520. See the Five Letters of Hernan Cortes.

5. See Robert I. Rotberg, *Health and Disease in Human History: A Journal of Interdisciplinary History Reader* (MIT Press, 2000), 168–171 for a discussion of the debate on the number of victims in Tenochtitlan.

6. Cuitlahuac was succeded as Tlatoani by Cuauhtemoc, a man less than 25 years old.

7. The Spanish undoubtedly appreciated the effects of disease upon a siege. A generation before in 1489, a Spanish army had been practically wiped out by typhus, losing 17,000 men during the siege of Granada.

8   Bernal Diaz del Castillo, *The True History of the Conquest of New Spain* (Hackett, 2012), 434.

9   Gonzalo Fernandez de Oviedo y Valdes, *Natural History of the West Indies,* Book IV, Chapter II.

10 Noble David Cook in *Demographic Collapse: Indian Peru, 1520–1620* (Cambridge University Press, 1982) estimates that the population of the empire was between 4 and 14 million at the time of its conquest, which would make it as large as Spain.

11 See Jose Toribio Pollo and James B. Kiracofe or Robert McCaa et al., "Why Blame Smallpox? The Death of the Inca Huayna Capac and the Demographic Destruction of Tawantinsuyu," *American Historical Association Annual Meeting* (Jan. 8–11, 2004) for a discussion of the various diseases which are held responsible for the death of Huayna Capac and others.

12 De Soto himself would succumb to fever while traveling through the region and was buried by his men in the Mississippi River in secret so as not to allow the Natives to realize that he was mortal.

13 Theories attributing the mounds to various groups from the Vikings to the Lost Tribes of Israel would be first attacked in the 1890's by Cyrus Thomas.

14 See H. Krauss, *Zoonoses: Infectious Diseases Transmissible from Animal to Humans* for an overall discussion of the process and various historical illnesses.

15 Syphilis has historically been included in the Columbian Exchange as originating in the Americas and arriving in Europe with the return of Spanish sailors to Naples in 1494. However, research over the past two decades has identified numerous prospective Syphilitic cases from ancient Pompeii and medieval England. Yet this does not necessarily negate the possibility that it existed in the Americas as well.

16 Jordi Gomez i Prat and Sheila Mendoca de Souza have done much research into the presence of Tuberculosis in both North and South America before the arrival of the Spanish. In some cases outbreaks of the disease reached epidemic scale in isolated settings due to a variety of population stresses and outside factors. These include over 133 confirmed cases stretching back almost 2000 years.

17 This would include the various parasites that undoubtedly populated the Americas before the arrival of Europeans including *giardia*.

18 Recent archaeological evidence points to the presence of lice in preserved Andean mummies and details of a typhus type disease from Aztec writings and oral histories.

19 Guzman Poma de Ayala, *Letter to a King*, 42.

20 Cocoliztli is the Nahuatl world for plague or pestilence.

21 Rodolfo Acuna-Soto, Leticia Calderon Romero, and James H. Maguire, "Large Epidemics of Hemorrhagic Fevers in Mexico, 1545–1815," *American Journal of Tropical Medicine and Hygiene* 62 (2000): 733–739.

22 Including Bolivian Hemorrhagic Fever and Argentine Hemorrhagic Fever.

23 R.C. Robertson, *Rotting Face: Smallpox and the American Indian* (Caldwell: Cauton Press, 2001), 52.

24 Henry Spelman. *Relation of Virginia* (Chiswick Press, 1872), 40. Cabeza de Vaca, *Chronicles of the Narvaez Expedition* (NY: W.W. Norton and Company, 2013), 35. Thomas Harriot would report a similar practice in his *Brief and True Report of the New Found Land of Virginia*.

CHAPTER 2

1   Duke of Medina-Sidona, *Letter to the King of Spain, Philip II* (June 24, 1588). English sources report that after the capture of the *Nuestra Senora del Rosario*, its bread was found to be "full of worms."
2   Letter to King Philip II (Sept, 1588).
3   Diary of Juan de Saavedra (Sept. 23, 1588).
4   Letter, Medina-Sidonia to Philip II (June 24, 1588).
5   Michael Lewis, *The Spanish Armada* (New York: Pan Books, 1960), 179.
6   Letter, Lord Howard to William Cecil (Aug. 20, 1588).
7   Drake suffered a similar fate in 1585. At the start of the Spanish War, Drake had sailed with 25 ships and 2300 men to raid Spanish settlements. Yet a week after docking at Cape Verde, 200 to 300 of his men would be killed by smallpox and his expedition would grind to a halt.

CHAPTER 3

1   Thomas Hariot, *A Brief and True Report of the New Found Land of Virginia* (1588).
2   Michael Leroy Oberg, *Dominion and Civility: English Imperialism and Native America 1595–1685* (Cornell University Press, 2004), 43.
3   See Peter B. Mires "Contact and Contagion: The Ronaoke Colony and Influenza," *Historical Archaelogoy*, Vol. 28 No 3 (1994).
4   John Smith, *The General History of Virginia.*
5   Stahle et al. 1998.
6   The founder of Cuttyhunk, Bartholomew Gosnold who also explored the Mass. Bay Region, would die of scurvy and dysentery in 1607 three months after being chosen to lead the new colony. See Charles Manning and Merrill Moore. "Sassafras and Syphillis." *The New England Quarterly* Vol. 9, No. 3 (Sept. 1936), 473–475.
7   Not a very popular member of the Pilgrim community, due to his arrogance and financial mismanagement, his death was hardly mourned.
8   G.F. Willison, *Saints and Strangers* (New York: Reynal and Hitchcock, 1945), 179.
9   Robertson, 105–106.
10 John Winthrop.
11 John S. Marr, "New Hypothesis for Cause of Epidemic Among Native Americans, New England 1616–1619," *Historical Review,* Vol. 16, No. 2 (Feb. 2010), http://wwwnc.cdc.gov/eid/article/16/2/09-0276_article.htm
12 Squanto was recorded as having once told the local Indians that the English controlled the plagues descending upon them. Keeping the disease, "buried under the storehouse." When questioned about this the English replied that, "the God of the English has it in store, and could send it at his pleasure, to the destruction of his or our enemies." Alfred W. Crosby, *Germs, Seeds, and Animals: Studies in Ecological History* (New York: Routledge, 1994), 116.
13 William Bradford, *Bradford's History of the English Settlement.*
14 Journals of Bradford and Winslow (March 16, 1621).
15 Cushman's Discourse, (Dec. 1621), 4.
16 Martha Lamb, *History of the City of New York: Its Origin, Rise, and Progress,* Volume I (Cosmo Inc., 2005), 200.
17 Richard W. Cogley, *John Elliot's Mission to the Indians Before King Philip's War* (Harvard

University Press, 2009), 132.

18 Francis Jennings, *The Invasion of America: Indians, Colonialism, and the Cant of Conquest* (New York: Norton, 1975), 15 and 26.

19 Daniel R. Mandell, *King Philip's War: The Conflict Over New England* (Infobase Publishing, 2009), 99.

20 See James David Drake, *King Philip's War: Civil War in New England* (University of Massachusetts Press, 1999), 168–170 for a discussion of casualty figures.

CHAPTER 4

1 Henry S. Burrage, ed. *Early English and French Voyages, Chiefly from Hakluyt, 1534–1608* (New York: Charles Scribner's Sons, 1906), 64.

2 Henry F. Dobyns, *Their Number Become Thinned—Native American Population Dynamics in Eastern North America* (Knoxville: University of Tennessee Press, 1983), 22 and Bruce Trigger, *Natives and Newcomers: Canada's Heroic Age Reconsidered* (Toronto: McGill Queen's University Press, 1986), 230.

3 Edna Kenton, *Jesuit Relations and Allied Documents: Travels and Explorations of the Jesuit Missionaries in New France* (Cleveland: Burrows Press, 1898).

4 Ibid., 10.

5 Letter from Jacques Bruyas (Jan.21, 1688) in *Jesuit Relations* 51:12.

6 Daniel K. Richter, "War and Culture: The Iroquois Experience," *The William and Mary Quarterly*, 3rd Ser., Vol. 40, No. 4. (Oct., 1983), 542.

7 Ibid, 534.

8 An outgrowth of these mourning wars would be the phenomenon of Captivity Narratives. Some of the first popular literature produced in America were stories of the horrors and adventures that befell the colonists taken during these raids. The Deerfield Massacre itself produced the captivity narrative of Eunice Kanenstenhawi Williams, the daughter of Minister John Williams who was adopted at the age of 7 by the Mohawks of Kahnawake Village. She would later marry a member of the tribe and despite pleas from her family would continue to live with the Mohawk until her death in 1785.

9 Jennifer Crump. *Canada Under Attack* (Toronto: Dundam Press, 2010), 19.

10 Letter, Fitz-John Winthrop to the Governor and Council of Connecticut (July 29, 1690).

11 Letter, Fitz-John Winthrop to the Governor and Council of Connecticut (Aug. 15, 1690).

12 Fitz-John Winthrop, "Journal of the Expedition to Canada."

13 John Wise, *Two narratives of the expedition against Quebec, A.D. 1690, under Sir William Phips*, 35–36.

14 Ibid., 38.

15 Walter Kendall Watkins, *Soldiers in the Expedition to Canada in 1690 and Grantees of the Canada Townships* (1898), 3.

16 Cotton Mather, *Magnalia Christi Americana*, 677.

17 Vernon's accomplishments were celebrated with the new British song, "Rule Britannia."

18 George Anson, *A Voyage Round the World* (New York: E.P. Dutton and Company), 16.

19 Ibid., 47.

20 Letter, James Lind to Thomas Pennant (Aug. 1793).

21 James Lind, "Two Papers on Fevers and Infection: which were read before the philosophical and medical society in Edinburgh" (1763).

22 The initial planned assault on Havana had been cancelled due to both its strength and the death of the main proponent for an assault on the city, Maj. Gen. Sir Charles Cathcart at sea while in route to the Caribbean.

23 Spotswood had died in Annapolis at the age of 64 prior to departing for the Caribbean.

24 Jonathan R. Dull, *The Age of the Ship of the Line: The British and French Navies, 1650–1815* (Lincoln: University of Nebraska Press, 2009), 46.

25 The choice of Lawrence Washington to command the colonial marines on board Vernon's flagship probably saved his life as conditions were infinitely better aboard the *HMS Princess Caroline*. Upon his return to Virginia, young Washington would settle down, purchase a plot of land, and name his new estate Mount Vernon in honor of his still revered commanding officer.

CHAPTER 5

1 Letter, Duquesne to Ministry (Nov. 29, 1753).

2 Letter, George Washington to John Augustine Washington (June 28, 1755).

3 Webb's 48th Regiment lost its medicine chest at the Monongahela and would spend two years without it during various campaigns against the French.

4 Washington continued to suffer from dysentery for many years, experiencing reoccurrences in 1757 and 1761. In the latter year the future president would contract malaria as well.

5 Quoted in Alfred Cave, *The French and Indian War* (Greenwood Publishing Group, 2004), 15.

6 Kenneth F. Kiple et al., *Biological Consequences of the European Expansion* (New York: Ashgate, 1997), 336.

7 William H. Fowler, *Empires at War: The French and Indian War and the Struggle for North America* (Bloomsbury Publishing, 2009), 137–139.

8 Gov. Shirley had previously quarantined a group of Acadians who had arrived by ship at Boston in December of 1755. They were refused permission to disembark for three months, causing the deaths of hundreds aboard due to disease.

9 Gen. Amherst had employed the same tactic upon taking Fort Ticonderoga earlier in the year, setting up breweries immediately following the fort's capture.

10 David Dixon, *Never Come to Peace Again: Pontiac's Uprising and the Fate of the British Empire in North America* (University of Oklahoma Press, 2005), 151.

11 Ensign Homes apparently was lured out by his Indian lover who requested his help in bleeding a gravely ill fellow native.

12 *Diary of William Trent* (June 24, 1763).

13 Smallpox had erupted in Boston in 1763 killing off 62% of the Nantucket Indians. That same year it would spread to South Carolina, Louisiana, and Pennsylvania.

14 Many historians have speculated that the First Fleet, which departed London for Australia only a quarter century later in 1787, purposefully spread smallpox among the Aborigines upon landing.

15 *Alexander Henry's Travels and Adventures in the Years 1760–1776* (Chicago: R.R. Donnelly and Sons, 1921), 237.

16 George Morgan, *Journal of George Morgan* (Nov. 28, 1766).

CHAPTER 6

1 P.H. Wood, "Sickness and settlement: disease as factor in early colonization of New England," *The Pharos of Alpha Omega Alpha* 27:98–101 (1964).

2 *A Report of the Record Commissioners of the City of Boston: Containing the Selectmen's Minutes from 1764 to 1768* (Boston: Heritage Books, 1889), contains hundreds of references to smallpox at almost every meeting during the epidemic.

3 Ibid., 39.

4 Esther Forbes, *Paul Revere* (New York: Houghton Mifflin, 1999), 76–78.

5 *A Report of the Record Commissioners of the City of Boston,* 23.

6 Jonathan R. T. Davidson, *Downing Street Blues: A History of Depression and Other Mental Afflictions in British Prime Ministers* (New York: McFarland, 2011), 22.

7 The American Stamp Act Congress itself was plagued by delay due to various smallpox outbreaks that kept a number of colonial legislatures at bay. Most notable among this list was the Assembly of Maryland, which had been prorogued in the spring due to an outbreak of the pestilence.

8 Davidson, 34–35.

9 See MacAlpine and Hunter, "Porphyria in the Royal Houses of Stuart, Hanover and Prussia," *British Medical Journal* 1, 5583 (Jan. 6, 1968), 7–18.

10 This was the same James' Powder commented upon by George Washington during the French and Indian War.

11 "Committee of Supplies directed to procure ten tons of Brimstone, and all kinds of Warlike Stores, sufficient for an Army of fifteen thousand Men to take the field, The Powder now at Concord, to be removed to Leicester." (Feb. 21, 1775) *American Archives* Series 4, Volume 1, p. 1367.

12 "Receiver General to pay to Doctor Warren and Doctor Church, five hundred Pounds, for the purchase of such articles for the Provincial Chests of Medicine, as cannot be got on credit. "*American Archives* Series 4, Volume 1, p. 1369.

13 Dr. Joseph Warren helped to treat victims of the attack upon the home of Ebenezer Richardson two weeks before the Boston Massacre . Following the attack he saved the life of Samuel Gore, pulling two bullets from him, and patching him up enough to allow him to participate in the Boston Tea Party. Warren also performed the official autopsy on Christopher Seider, whose findings were used to convict Richardson of murder.

14 Letter, George Washington to President of Congress (July 21, 1775).

15 Letter, George Washington to John Augustine Washington (July 27, 1775).

16 Charles Martyn, *The Life of Artemas Ward, the First Commander-in-Chief of the American Revolution* (Boston: A. Ward, 1921), 165.

17 Letter, George Washington to General Gage (August 11, 1775).

18 Letter, Robert H. Harrison to Council of Massachusetts (December 3, 1775).

19 Letter, George Washington to President of Congress (December 19, 1775).

20 Letter, George Washington to Joseph Reed.

21 John E. Ferling, *The First of Men: A Life of George Washington* (New York: Oxford

University Press, 2010), 139.

22 The hero of the action, Ethan Allen was once brought to court and fined for inoculating himself against smallpox.

23 Letter, George Washington to John Hancock (March 19, 1776).

24 David J. Kiracofe, "Dr. Benjamin Church and the Dilemma of Treason in Revolutionary Massachusetts," *New England Quarterly,* 70 (1997), 443–462.

25 Benjamin Rush, *The Autobiography of Benjamin Rush* (Princeton: Princeton University Press, 1948), 131–132.

26 Letter, Benjamin Rush to George Washington (Dec. 26, 1777).

27 As quoted in *Proceedings of the Connecticut Medical Society* (Connecticut Medical Society, 1867), 302.

28 Letter, Dr. Benjamin Rush (April 22, 1777).

29 James Tilton, *Economical Observations on Military Hospitals: And the Prevention and Care of Diseases Incident to an Army* (Wilmington: 1813), 15.

30 Isaac Senter. "The Journal of Isaac Senter, Physician and Surgeon to the Troops Detached from the American Army Camped at Cambridge, Mass., on a Secret Expedition Against Quebec Under the Command of Col. Benedict Arnold, in September, 1775." (Sept. 27, 1775).

31 Ibid. (Oct. 16th and 17th).

32 *The Diary of Caleb Haskell* (Newburyport: W.H. Huse and Company, 1881), 14.

33 Letter, Thomas Jefferson to Francois Soules.

34 Charles Coffin, *The Life and Services of Major Gen. John Thomas* (NY: Egbert, Hovey, and King, 1844), 25.

35 Isaac Senter.

36 Coffin, 29–30.

37 Letter, John Sullivan to Philip Schuyler (June 19, 1776).

38 Letter, John Adams to Abigail Adams (July 7, 1776).

39 Letter, George Washington to John Hancock (July 14, 1776).

40 Letter, John Adams to Abigail Adams (June 26, 1776).

41 Letter, Jose de Galvez to Luis de Unzaga (Dec. 24, 1776).

42 Thurston (1940) as summarized in Donald R. Hopkins, *The Greatest Killer: Smallpox in History* (Chicago: University of Chicago Press, 2002), 73.

43 Ibid., 73–74.

44 Save for around 300 sick men who were left behind at the hospital in Boston. From an April 3rd Letter from Gen. Washington to Dr. John Morgan.

45 Dr. James Tilton, as quoted in Louis C. Duncan, "Medical Men in the American Revolution: The New York Campaign of 1776." *New York Medical Journal* Vol. CXII No. 11 (Sept. 11, 1920), 1.

46 Charles H. Lesser. "The Sinews of Industry: Monthly Strength Reports of the Continental Army" (Chicago 1976) xxx, xxi.

47 David Hackett Fischer, *Washington's Crossing* (New York: Oxford University Press, 2006), 86.

48 Letter, General Lee to Congress (February 1776).

49 Bruce Bliven Jr., *Under the Guns* (New York: Harper and Row, 1972).

50 The Hessian's use of sauerkraut helped to mitigate the occurrence of scorbutic diseases among their units.

51 "Popp's Journal," *Pennsylvania Magazine of History and Biography*, Vol. 26 (1902)
52 Nathanael Greene as recounted in Theodore Thayer, *Nathanael Greene: Strategist of the American Revolution* (New York: Twayne, 1960), 95.
53 Letter, George Washington to John Hancock (Sept. 8, 1776).
54 Letter, George Washington to John Adams (Sept. 22, 1776).
55 Letter, John Adams to Abigail Adams (April 13, 1777).
56 Letter , Lord Dunmore to Secretary of State (July 31, 1776).
57 *Virginia Gazette* (June 15, 1776).
58 Letter, Lord Dunmore (Jan. 16, 1776).
59 Letter, Archibald Cary to R. H. Lee (December 24, 1775).
60 Letter, Richard Hutson to Isaac Hayne in *Richard Hutson Letterbook*, Langdon Cheves III Papers: Extracts from Private Journals (May 27, 1776).
61 Henry Clinton, *The American Rebellion.*, ed. William Willcox (New Haven: Yale University Press, 1954) 26–29.
62 Quoted in Rufus R. Wilson, *New York: Old & New; Its Story, Streets, and Landmarks* Vol. 1 (New York: 1903), 236–237.
63 From Robert Sheffield as quoted in John Warner Barber, *Connecticut Historical Collections, Containing a General Collection of Interesting Facts, Traditions, Biographical Sketches, Anecdotes, Etc. Relating to the History and Antiquities of Every Town in Connecticut* (Durrie and Peck, 1945), 286–287.
64 As reported in the Boston Gazette in August of 1781.
65 *An Historical Sketch, to the End of the Revolutionary War, of the Life of Silas Talbot* (New York: G&R Waite, 1803), 107.
66 *New Hampshire Gazette* (April 26, 1777).
67 *The Newark Advocate* (Newark, Ohio) (Sept. 19, 1912).
68 David Swain, *Recollections of the Jersey Prison Ship*, ed. Thomas Dring (Westholme Publishing, 2010). Dring's original work was published in 1829.
69 Letter, George Washington to Sir William Howe (January 13, 1777).
70 Benjamin DeWitt, *An Account of the Internment of the Remains of 11,500 American Seamen, Soldiers, and Citizens, Who Fell Victims to the Cruelties of the British on board their Prison Ships at the Wallabout During the American Revolution* (1808).
71 Robert Donkin, *Military Collections and Remarks* (New York: H. Gaines, 1777).
72 Letter, George Washington to William Shippen (January 6, 1777).
73 The operation only took in one child, with the other needing three attempts at a good amount of wine to speed along the process.
74 Johann David Schoepf, *Climate and Diseases of America* (Boston: H.O. Houghton & Co., 1875).
75 George Washington as quoted by Shaul G. Massry et al., "History of Nephrology," *The American Journal of Nephrology* 17 (1997), 233–240.
76 Tilton, 27–29.
77 Ibid.
78 Jones was the attending physician at the time of Benjamin Franklin's death and helped to save Pres. Washington's life in 1790. He would later pass away from an illness he received visiting Washington in 1791.
79 Tilton, 27–29.

80 Peter Whiteley, *Lord North: The Prime Minister who Lost America* (A&C Black, 1996), 189.

81 Ibid., 166.

82 Letter, George Washington to Landon Carter (May 30, 1778).

83 John Montresor, *The Montresor Journals* Vol. 14 (1882), 507.

84 Letter, Augustine Prevost to Sir Henry Clinton (July 14 and July 30, 1779).

85 Lincoln as quoted in Peter McCandless, *Slavery, Disease and Suffering in the Southern Lowcountry* (New York: Cambridge University Press, 2014), 88.

86 One of the British dead was John Maitland, a very talented officer responsible for Prevost's victory at Stono Ferry the previous spring, who succumbed to fever shortly after the siege. One of the American dead was Casimir Pulaski, who was killed by grapeshot while rallying the French cavalry. Both deaths were heavy blows to their respective sides.

87 Letter, Lord Cornwallis to Clinton (Aug. 20, 1780).

88 Letter, Lord Cornwallis to Lord George Germain (Aug. 20, 1780).

89 McCandless, 95–97.

90 Letter, Lord Cornwallis to Lord George Germain (Sept. 19, 1780).

91 Letter, Lord Cornwallis to Lt. Col. Turnbull (Oct. 7, 1780).

92 Charles Ross, *The Correspondence of Charles, First Marquis Cornwallis* Vol. 1 (London: 1859), 59 footnote.

93 Letter, Josiah Smith to James Poyas (Dec. 5, 1780), in *Josiah Smith Letter Book, 1771–1784*, 411.

94 Letter, Lord Cornwallis to Clinton (April 10, 1781) This letter is also in Henry Clinton, *The American Rebellion,* ed. by William B. Willcox (New Haven: Yale University Press, 1954), 508–10.

95 Schoepff, 7.

96 Gen. Alexander Leslie proposed a similar plan to Cornwallis, expressing his desire to send 700 infected escaped slaves up the river towards various rebel plantations.

97 Letter, Thomas Jefferson to William Gordon (July 16, 1788).

98 Benson Bobrick, *Angel in the Whirlwind.* Though the number has been questioned by many (see Cassandra Pybus. "Jefferson's Faulty Math: The Question of Slave Defections in the American Revolution," *The William and Mary Quarterly.* Third Series, Vol. 62, No. 2 . (April 2005), 243–264), nonetheless defection and disease did go hand-in-hand and killed thousands at least.

99 Benjamin Franklin, *The Retort Courteous* (1786).

100 Letter, Benjamin Franklin to David Hartley (Sept. 17, 1782).

101 Some of the more notable people included on this list would be John Parke Custis, Washington's stepson who succumbed to typhoid after Yorktown, Col. Robert Erskine, the grand cartographer of the war and inventor of the chain that stretched across the Hudson at West Point, who died of pneumonia in 1780, and Gen. Enoch Poor who succumbed to typhus in 1780.

102 See Elizabeth Fenn's *Pox Americana.*

CHAPTER 7

1 Letter, James Madison to James Madison Sr. (April 1, 1787).

2   *Boston Gazette* (Aug. 15, 1774).

3   Alden Bradford, *History of Massachusetts* (Richardson and Lord, 1825), 19.

4   Letter, John Quincy Adams to Abigail Adams (Aug. 27, 1785).

5   Letter, Henry Knox to George Washington (Oct. 23, 1786).

6   Little Turtle would be inoculated at some point in the 1790s against smallpox by Dr. Benjamin Rush. It is unclear whether this was due to a practical or tactical desire to avoid the deadly contagion.

7   Letter, John Hamtramck to Josiah Harmar (July 29, 1789).

8   Celia Barnes, *Native American Power in the United States, 1783–1795* (Madison: FDU Press, 2013), 133.

9   Letter, Arthur St. Clair to George Washington (November 9, 1791).

10 Letter, R. G. England to John Graves Simcoe in *Simcoe Papers,* Vol. III, 22.

11 "Daily Journal of Wayne's Campaign," (Sept. 2, 1794) in Samuel Prescott Hildreth, *Original Contributions to the American Pioneer* (Cincinnati: John S. Williams, 1814), 351.

12 Brenda Gayle Plummer, *Haiti and the United States: Psychological Movement* (University of Georgia Press, 1992), 18–19.

13 Thomas O. Ott, *The Haitian Revolution, 1789–1804* (University of Tennessee Press, 1973), 146.

14 Letter, Charles Leclerc to Napoleon (Oct. 2, 1802).

15 Jon Kukla, *A Wilderness So Immense* (New York: Alfred A. Knopf, 2003), 223.

16 Howard N. Simpson, *Invisible Armies: The Impact of Disease on American History* (Indianapolis: Bobbs-Merrill Company, 1980), 166.

17 Barnett Singer et al., *Cultured Force: Makers and Defenders of the French Colonial Empire* (Madison: University of Wisconsin Press, 2008), 45.

18 Jefferson was described by William J. Hyland in 1808 as, "*a frail sixty-four years old, suffering from excruciating migraine headaches, debilitating rheumatoid arthritis, diarrhea, and numerous intestinal infections.*"

19 Julie Fanselow, *Traveling Lewis and Clark Trail* (Guilford, CT: Globe Pequot Press, 2007), 203.

20 One of the theories surrounding the death of Meriwether Lewis was that his use of mercury on the expedition led to mental illness and his subsequent suicide in 1809.

21 Meriwether Lewis, *Journal of Meriwether Lewis* (Jan. 27, 1806).

22 Ibid. (June 10, 1805).

23 See Ronald V. Loge, M.D., 'Two dozes of barks and opium': Lewis and Clark as Physicians," *We Proceeded On,* Vol. 23, No. 1 (February 1997), 10–15, 30. Reprinted from *The Pharos of Alpha Omega Alpha,* Vol. 59, No. 3 (Summer, 1996), 26–31.

24 Sacagawea herself would die from "*putrid fever*" in 1812, largely forgotten, at Fort Lisa in North Omaha, Nebraska.

25 Ballard C. Campbell, *Disasters, Accidents, and Crises in American History* (Infobase Publishing, 2008), 53.

26 Letter, Thomas Jefferson to Dupont de Nemours (July 1807) in *The Jefferson Cyclopedia* (Funk and Wagnells, 1907), 137.

27 These included James McHenry (1796–1800), Henry Dearborn (1801–1809), and William Eustis (1809–1813).

28 Letter, Dr. Benjamin Rush to John Adams (July 4, 1798).

29 Simpson, 103.

30 James Mann, *Medical sketches of the campaigns of 1812, 13, 14* (H. Mann & Co, 1816), vi.

31 Ibid, 14–16.

32 John Douglas, *Medical Topography of Upper Canada* (London: Burgis and Hill, 1819), 27.

33 Testimony of Dr. W.M. Ross (Fall 1813) in James Wilkinson's *Memoirs of My Own Times*, 3 Vols (Philadelphia, 1816), 3:111, p 308 quoted in Donald Hickey, *The War of 1812* (Chicago: University of Illinois Press, 1989), 78–79.

34 Mann, 24.

35 Ibid, 30.

36 Carl Edward Skeen, *Citizen Soldiers in the War of 1812* (University Press of Kentucky, 1999), 54.

37 Mann, 125.

38 John K. Mahon, *The War of 1812* (Gainesville: University of Florida Press, 1972), 82.

39 Alan Taylor, *The Civil War of 1812* (NY: Alfred A. Knopf, 2010), 193.

40 Simpson, 107.

41 Letter, Benjamin Rush to John Adams (March 16, 1813).

42 Letter, James Madison to Henry Dearborn (August 8, 1813). Dearborn's ineffectiveness probably also played a large role in this.

43 Mahon, 171.

44 Gen. Wilkinson had achieved notoriety for not only being involved in a plot with Vice President Aaron Burr to seize half of the United States, but had been, when in charge of American forces in New Orleans before the war so incompetent in the supplying of his men's needs, that while he lived in luxury over 1000 of them (half his force) had died from disease or fled due to its presence.

45 Mann, 119.

46 Ibid.

47 Robert Sherman Quimby, *The US Army in the War of 1812: An Operational and Command Study* (Michigan: Michigan University Press, 1997), 311.

48 Mann, 173.

49 Ibid, 44.

50 Taylor, 198.

51 *Centinel* (Dec. 17, 1813).

52 Hickey, 302.

53 Marion Breunig, "A Tale of Two Cities: Washington and Baltimore During the War of 1812," in *War in an Age of Revolution* (New York: Cambridge University Press, 2010), 353–372.

CHAPTER 8

1 John Dominis would go on to achieve fame as both the father-in-law of Queen Liliuokalani of Hawaii and for disappearing under mysterious circumstances on a voyage to China in 1846.

2 Jim Mockford, "The Mystery of the Brig *Owhyhee*'s Anchor and the Disappearance of Captain John Dominis," *The Northern Mariner* XVIII, Nos. 3–4 (July–Oct. 2008), 113.

3   Robert H. Ruby, *The Chinook Indians: Traders of the Lower Columbia River* (University of Oklahoma Press, 1987), 185–190

4   W.G. Kendrew, *The Climates of the Continents* (Oxford: Clarendon Press, 1937).

5   Only 14 years after their own demographic disaster, the Choctaw famously sent $710 that they had raised to Ireland in 1847 to help provide relief for those suffering from the potato blight.

6   Private John G. Burnett of Captain Abraham McClellan's Company,2nd Regiment, 2nd Brigade, Mounted Infantry recorded his memories about the Cherokee Indian Removal during 1838–39 in a letter to his children (Dec. 11, 1890) accessible via http://www.digitalhistory.uh.edu/disp_textbook.cfm?smtID=3&psid=1147

7   George McClellan of Civil War fame was stricken with both dysentery and malaria during the war and battled reoccurring bouts of the latter for years afterwards.

8   Included among this number are Stephen Kearny, the conqueror of Santa Fe and California, who contracted malaria or yellow fever and died a year after his return. Seymour V. Connor, *North America Divided: The Mexican War 1846–1848* (NY: Oxford Press, 1971), 171.

9   K.J. Bauer, *The Mexican War* (New York: Macmillan, 1974), 206 and 209.

10 Ulysses S. Grant, *Personal Memoires of US Grant* (Gutenberg eBook, 2004), 54. Grant's own commander died of heart disease on his first day at Camp Salubrity in New Orleans. See Grant, 21.

11 *Martinsburg Gazette* (Feb. 18, 1847).

12 *Nile's National Register* (Feb. 6, 1847).

13 *Martinsburg Gazette* (March 23, 1847).

14 Merrill Mattes, *The Great Platte River Road* (Lincoln: University of Nebraska Press, 1987), 82 and 85.

15 Peter D. Olch, "Treading the Elephant's Tail: Medical Problems on the Overland Trails." *Overland Journal*, Vol 6, Num 1 (1988): 25–31.

16 Katherine E. Krohn, *Wild West Women* (New York: Lerner, 2005), 26.

17 Grant, 85.

18 Ibid.

19 See Grant, 88–89 for a discussion of his observations of these practices during the outbreak.

20 Calhoun's speeches on the matter had to be read by various other members of Congress as he was too sick to attend sessions.

21 Arnold Krupat, *'That the People Might Live': Loss and Renewal in Native American Elegy* (Cornell University Press, 2012), 46.

CHAPTER 9

1   Horatio King, *Turning on the Light: A Dispassionate Survey of President Buchanan Administration* (J.B. Lippincott, 1895), 192.

2   The total number of deaths in the war is still a debated point today, see Guy Gugliotta's article "New Estimate Raises Civil War Death Toll," in the *NY Times* on April 2, 2012 for a discussion of the most recent research.

3   H.H. Cunningham, *Doctors in Gray: The Confederate Medical Service* (Gloucester, Mass: Peter Smith, 1970), 185.

4   Thomas Power Lowry, *The Stories the Soldiers Wouldn't Tell: Sex in the Civil War* (Stack-

pole Books, 2012), 104.

5   "African-American Civilians and Soldiers Treated at Claremont Smallpox Hospital, Fairfax County, Virginia, 1862–1865," http://www.freedmenscemetery.org/resources/documents/claremont.pdf (Accessed April 1, 2015).

6   Letter, Charles Furman to Fannie Garden Furman (Nov. 1861).

7   William C. Davis, *Battle at Bull Run: A History of the First Major Campaign of the Civil War* (Louisiana State University Press, 1981), 57.

8   Terry L. Jones, "Brother Against Microbe," *NY Times* (Oct. 26, 2012).

9   Daniel O'Flaherty, *General Jo Shelby: Undefeated Rebel* (UNC Press Books, 2000), 128.

10 William Mathias Lamers, *The Edge of Glory: A Biography of William S. Rosecrans* (LSU Press, 1999), 62.

11 Rick Steelhammer, "Robert E. Lee: 'Outwitted, outmaneuvered, and outgeneraled,'" *Charleston Gazette* (Oct. 29, 2011).

12 *New York Tribune* (Dec. 31, 1861).

13 Letter, George Meade to his wife (Jan. 5, 1862) in George Gordon Meade, *Life and Letters* Vol. 1 (New York: Charles Scribner's Sons, 1913), 242.

14 Cunningham, 191.

15 Letter, J. C. Nott to Joseph Jones (Nov. 21, 1866) in Joseph Jones, *Contagious and Infectious Diseases* (L. Jastremski, 1887), 327.

16 Chester G. Hearn, *Lincoln and McClellan at War* (LSU Press, 2012), 133.

17 Grant, 205.

18 Ibid.

19 Robert K. Krick, "The Army of Northern Virginia in September 1862," 42 in Gary W. Gallagher, ed. *Antietam: Essays on the 1862 Maryland Campaign* (Kent State University Press, 1989).

20 Brenda Chambers McKean, *Blood and War at my Doorstep* Vol. 2 (Xlibris Corporation, 2011), 901.

21 J.G. Westmorland et al. *Atlanta Medical and Surgical Journal* (1867), 242.

22 See Michael Fitzgerald's "Did 'Stonewall' Jackson Have Asperger's Syndrome?" *Society of Clinical Psychologists* Accessed at http://www.scpnet.com/paper2_2.htm

23 See Beverly C. Smith, "The Last Illness and Death of Thomas Jonathan (Stonewall) Jackson," *VMI Annual Review* (1975), accessed at http://www.vmi.edu/uploaded-Files/Archives/Jackson/Death_and_Funeral/The%20Last%20Illness%20and%20Death%20of%20General%20Jackson.pdf

24 Letters, Robert E. Lee to Mary Lee (April 11, 1863) in *War Papers of Robert E. Lee* (Boston: Little Brown & Co, 1961).

25 R.D. Mainwaring et al., "The Cardiac Illness of Robert E. Lee," *Surg. Gynecol. Obstet.* Vol. 174, No. 3 (March 1992), 237–244.

26 Dr. Carl Coppolino, "Lee's Illness Lost Gettysburg," *The Gettysburg Magazine*, No. 46 (Jan. 2011).

27 Letter, Robert E. Lee to Jefferson Davis (Aug. 8, 1863).

28 Pres. Lincoln himself would contract smallpox shortly after giving the Gettysburg Address and was incapacitated in the White House from November to January.

29 Hill's contraction of the disease caused him to have to repeat his 3rd year at the military institute.

30 Gene C. Armistead, *Horses and Mules in the Civil War* (McFarland, 2013), 62.
31 R. Douglas Hurt, *Agriculture and the Confederacy: Policy, Productivity, and Power in the Civil War South* (UNC Press Books, 2015), 141.
32 New Orleans Commercial Bulletin (Sept. 26, 1867).
33 William A. Hammond, "Circular No. 6," (May 4, 1863).

CHAPTER 10

1 U.S. Marine Hospital Service. *Annual Report of the Supervising Surgeon General of the Marine Hospital.* (US Government Printing Office, 1896), 388.
2 Ibid, 389.
3 *Manufacturers and Farmers Journal* (June 2, 1898).
4 *Morning Record* (May 1, 1898).
5 *The Evening Record* (April 22, 1898).
6 *Sunday Herald* (May 10, 1898).
7 *NY Times* (May 4, 1898).
8 "To Ward Off Yellow Fever," *Kansas City Journal* (June 14, 1898), 1.
9 Theodore Roosevelt, *The Rough Riders* (New York: Library of America, 2004), Chapter III, 52.
10 Ibid, 140.
11 Letter, William R. Shafter to Woodrow Wilson (Feb. 9, 1899).
12 Ibid, Appendix C "Round Robin Letter."
13 See Martha L. Sternberg, *George Miller Sternberg: A Biography* (American Medical Association, 1920), 305–307.
14 Frederic H. Sawyer, *The Inhabitants of the Philippines* (New York: C. Scribner and Sons, 1900), 114.
15 Ken De Bevoise, *Agents of Apocalypse: Epidemic Disease in the Colonial Philippines* (Princeton University Press, 1995), 86.
16 Joseph R. Svinth, "The Origins of Philippines Boxing," *Journal of Combative Sport* (July 2001).
17 Bessie L. Scovell, Excerpts from the President's Address, "Minutes of the Twenty–Fourth Annual Meeting of the WCTU of the State of Minnesota," (1900).
18 Carol R. Byerly, *Good Tuberculosis Men: The Army Medical Department's Struggle with Tuberculosis* (Dept. of the Army, 2014), 7.
19 Ibid., 9.

CHAPTER 11

1 Jack London, "Yah, Yah, Yah," in *South Sea Tales.*
2 Jack London, "The Unparalleled Invasion."
3 "Biological and Chemical Warfare: An International Symposium," *Bulletin of the Atomic Scientist* (June 1960), 228.
4 W. Seth Carus, "Bioterrorism and Biocrimes: The Illicit Use of Biological Agents Since 1900." (Washington DC: Center for Counterproliferation Research, National Defense University, 1998), 297.
5 Anton Casimir Dilger would die in 1918 of the flu during the great Spanish Influenza outbreak.
6 See Robert Koenig's *The Fourth Horseman: One Man's Mission to Wage the Great War*

in America (Public Affairs, 2009) for an excellent account of the Dilger affair.

7 *Oil, Paint, and Drug Reporter*, Vol. 92, No. 6 (Aug. 6, 1917), 16.

8 N.A. Metcalfe, *A Short History of Biological Warfare: Medicine, Conflict, and Survival* (2002), 271–282 .

9 M. Hugh-Jones, "Wickham Steed and German Biological Warfare Research," *Intelligence and National Security*, Number 7 (1992), 379–402.

10 Sharad Chauhan, *Biological Weapons* (APH Publishing, 2004), 193.

11 Reprint from the Public Health Reports, Issues 285–326 (United States Public Health Service), 5.

12 Allan M. Brandt, "The Syphilis Epidemic and its Relation to AIDS," *Science* 239 (January 1988), 377.

13 Col. Leonard P. Ayers, "The War with Germany, a Statistical Summary," Chapter IX (Washington: Government Printing Office, 1919).

14 Ibid. and Yale Kneeland, "Respiratory Diseases," John Boyd Coates, ed. *Internal Medicine in World War II* Vol. II (Office of the Surgeon General, 1963), 1.

15 Estimates of total deaths vary by source from 25 million to over 100 million and are severely limited by war time censorship and lack of reporting from more remote regions and colonies.

16 Josie Brown, "A Winding Sheet and a Wooden Box," *Navy Medicine*, Vol. 77, No. 3 (1986), 18–19.

17 The village of Brevig, Alaska lost 90% of its Inuit population. Many of these were buried in hastily dug graves under feet of permafrost. Dug up in 1951 and again in 2001, their remains have provided much information on the makeup of the virus.

18 See John M. Barry's *The Great Influenza* for a deeper depiction of the effect of the disease upon Wilson.

19 Letter, Cary T. Grayson to Samuel Ross (April 14, 1919).

20 Letter, William H. Taft to A. L. Lowell (Oct. 5, 1919).

CHAPTER 12

1 The Golden Horde under Janibeg famously catapulted the bodies of Bubonic Plague victims into the Genoese controlled town of Caffa in 1347 to affect the port's surrender. Some historians consider this to be the touching off point for the spread of the Black Death into Europe in the following year.

2 "Infectious, or Catarrhal Jaundice," *Chicago Tribune* (March 5, 1947).

3 One of the most famous episodes of an outbreak of disease in a training camp in the United States involved the eruption of Ft. Bragg Fever in July to August of 1942, a strange disease characterized by the outbreak of a rash on the victim's body around the fourth day after infection.

4 William A. Reilly, "Sandfly Fever," in John Boyd Coates, ed. *Internal Medicine in World War II* Vol. II (Office of the Surgeon General, 1963), 7–8.

5 Annex No. 2 to G-3 Report , Torch Operation, Field Orders (Nov. 17, 1942).

6 William A. Reilly, 56.

7 Denis Avey, *The Man Who Broke into Auschwitz* (Da Capo Press, 2012), 62–63.

8 See Cigdem Alkan et al., "Sandfly-borne Phleboviruses of Eurasia and Africa: Epidemiology, Genetic Diversity, Geographic Range, and Control Measures," *Antiviral Research,* Vol. 100, No. 1 (Oct. 2013), 54–74 for a discussion of the effect of the

sandfly on Allied and Axis operations.

9   Rick Atkinson, *The Day of Battle: The War in Sicily and Italy, 1943–1944* (Macmillan, 2008), 147.

10 See F.L. Soper et al., "Typhus Fever in Italy, 1943–1945, and its Control with Louse Powder," *The American Journal of Hygiene*, Vol. 45, No. 3 (May 1947), 305–334, for an in-depth analysis of the epidemic and its treatment.

11 Frank Snowden, *The Conquest of Malaria: Italy, 1900–1962* (New Haven: Yale, 2008), 192–198.

12 Stephen Arnon et al. "Botulinum Toxin as a Biological Weapon," *JAMA* Vol. 285, No. 8 (Feb. 28, 2001), 1060.

13 John Hedley-Whyte and Debra R. Milamed, "Surgical Travellers: Tapestry to Bayeux," *Ulster Medical Journal*, 83 (3), 171–177.

14 Eric Croddy, "Tularemia, Biological Warfare, and the Battle for Stalingrad," *Military Medicine*, Vol. 166, No. 10 (Oct. 2001).

15 Of the roughly 100,000 German prisoners taken at Stalingrad, 40% would die of typhus and be buried in mass graves by the Soviets.

16 Barbara Brooks Tomblin, *G. I. Nightingales: The Army Nurse Corps in World War II* (University Press of Kentucky, 2003), 57.

17 See "The Control of Dengue on Saipan," *Bulletin of the US Army Medical Department*, No. 84 (Jan. 1945), 1.

18 *Preventative Medicine in World War II* (US Army Medical Service: Office of the Surgeon General, 1963), 507.

19 Stanley Sandler, ed. *World War II: An Encyclopedia* (New York: Routledge, 2003), 617.

20 "Out of the Air," *The Listener*, Vol. 92 (BBC, Sept. 12, 1974), 332. The story may be apocryphal as no mention has been found from the 1940s itself and may be a conflation of the British use of gin and tonic to combat tropical diseases.

21 Letter, Dr. Frank Lahey (July 10, 1944).

22 Patrick Hurley, *Atlantic Monthly* (Sept. 28, 1950).

23 The United States did experiment in the 1950s with releasing bacterial agents, especially against crops, by E77 balloon bombs in a similar way to the Japanese in the 1940s.

24 "Judge's Decision Expected Soon in California German Warfare Case," *NY Times* (April 15, 1981).

25 *Medical Aspects of Biological Warfare* (GPO: Dept. of Defense, 1997), 565–568.

26 Jon Mitchell, "Were we Marines Used as Guinea Pigs on Okinawa" *The Japan Times* (Dec. 4, 2012) accessed at http://www.japantimes.co.jp/community/2012/12/04/issues/were-we-marines-used-as-guinea-pigs-on-okinawa/#.UbU92iO9Wm4.

27 Telegram, Mao Zedong to I. V. Stalin (Feb. 21, 1952).

28 *Komsomolskaya Pravda* (May 27, 1992), 2.

29 William Anderson, "Probability Analysis for the Case of Yellow Rain," *International Journal of Intelligence and Counterintelligence*, Vol. 3 (Spring 1989), 77–82.

30 Beatrice A. Golomb, "A Review of the Scientific Literature As It Relates to Gulf War Illnesses," *Rand Corporation* (1999).

31 Table 7 from Andre Ognibene, *Internal Medicine in Vietnam* Vol. II (Washington: Office of the Surgeon General, 1982), 65.

32 John Parascandola, *Sex, Sin, and Science: A History of Syphilis in America* (New York: Greenwood, 2008), 143.

33 Ibid., 146.

CHAPTER 13

1   David R. Petriello, "Yellow Fever, Yellow Peril: Disease and Nativism in American History," presented before the SAHMS Conference in Jackson, MS (March 2015).
2   W. Seth Carus, "RISE: A Case Study," *Encyclopedia of Bioterrorism Defense* (John Wiley & Sons, 2005).
3   "Anthrax God's Work, Says Bin Laden," *Telegraph* (Feb. 2, 2002).
4   "US Germ Warfare Research Pushes Treat Limits," *NY Times* (Sept. 4, 2001).
5   "White House: CIA has Ended Use of Vaccine Programmes," *BBC* (May 20, 2014)
6   "Al-Qaeda's Bio Weapons," *CBS News* (Feb. 11, 2009).
7   "Al-Qaeda Cleric's Call From Grave: Attack With Bio Weapons," *ABC News* (May 3, 2012).
8   "Al-Qaeda Cell Killed by Black Death was 'Developing Biological Weapons," *The Telegraph* (Jan. 20, 2009).
9   "Al-Qaeda Eyes Bio Attack from Mexico," *Washington Times* (June 3, 2009).
10 Harold Doornbos, "Found: The Islamic State's Terror Laptop of Doom," *Foreign Policy* (Aug. 28, 2014).
11 Ibid.
12 Milton Leitenberg, "Assessing the Biological Weapons and Bioterrorism Threat," (Strategic Studies Institute, Dec. 2005).

BIBLIOGRAPHY

A., Addison. *Addison A. to Sanford Ferguson* (July 18, 1832), Letter. *The Ferguson-Jayne Papers, 1826–1938*, edited by Mary S. Briggs. Interlaken New York: Heart of the Lakes Publishing, 1981.

"A Brief Narrative of the Ravages of the British and Hessians at Princeton in 1776." (Princeton: Princeton University Press, 1906).

Ackerknecht, Erwin H. "Malaria in the Upper Mississippi Valley 1660–1900." Supplements of the History of Medicine, No. 4, viii. Baltimore: J. Hopkins Press, 1945.

Acuna-Soto, Rodolfo, etal. "Large Epidemics of Hemorrhagic Fevers in Mexico, 1545–1815," *American Journal of Tropical Medicine and Hygiene* 62 (2000) pp. 733–739.

Acuna-Soto, Rodolfo, et al. "Megadrought and Megadeath in 16th Century Mexico." Emerging Infectious Diseases. April 2002. http://wwwnc.cdc.gov/eid/article/8/4/01-0175.htm. Accessed October 1, 2012.

Alchon, Suzanne Austin. *A Pest in the Land: New World Epidemics in a Global Perspective*. Albuquerque: University of New Mexico Press, 2003.

Aldridge, Alfred Owen. *Benjamin Franklin: Philosopher and Man*. Lippincott, 1965.

Alkan, Cigdem et al. "Sandfly-borne Phleboviruses of Eurasia and Africa: Epidemiology, Genetic Diversity, Geographic Range, and Control Measures." *Antiviral Research,* Vol. 100, No. 1 (Oct. 2013), 54–74.

American Archives Series 4, Volume 1.

Anderson, Fred. *Crucible of War*. NY: Vintage Books, 2000.

Anderson, William. "Probability Analysis for the Case of Yellow Rain." *International Journal of Intelligence and Counterintelligence*, Vol. 3 (Spring 1989), 77–82.

Armistead, Gene C. *Horses and Mules in the Civil War*. McFarland, 2013.

Arnon, Stepehn et al. "Botulinum Toxin as a Biological Weapon." *JAMA* Vol. 285, No. 8 (Feb. 28, 2001), 1059–1068.

Atkinson, Rick. *The Day of Battle: The War in Sicily and Italy, 1943–1944*. Macmillan, 2008.

Avey, Denis. *The Man Who Broke into Auschwitz*. Da Capo Press, 2012.

Ayers, Col. Leonard P. "The War with Germany, a Statistical Summary," Chapter IX. Washington: Government Printing Office, 1919.

Baine, Rodney M. *The Prison Death of Robert Castell and its Effects on the Founding of Georgia*. The Georgia Historical Quarterly Vol. 73, No. 1 (Spring 1989).

Barber, John Warner. *Connecticut Historical Collections, Containing a General Collection of Interesting Facts, Traditions, Biographical Sketches, Anecdotes, Etc. Relating to the History and Antiquities of Every Town in Connecticut*. Durrie and Peck, 1945.

Barnes, Celia. *Native American Power in the United States, 1783–1795*. Madison: FDU Press, 2013.

Barry, John M. *The Great Influenza*. New York: Penguin Books, 2005.

Bilkas, G.A. *General John Glover and his Marblehead Mariners*. New York: Henry Holt & Company, 1960.

"Biological and Chemical Warfare: An International Symposium," *Bulletin of the Atomic Scientist* (June 1960), 228.

Blake, John B. "The Inoculation Controversy in Boston: 1721–1722." *The New England Quarterly* 25:4 (Dec. 1952), pp. 489–506.

Blakely, Robert L. and Philip L. Walker. "Mortality Profile of the Middle Mississippi Population of the Dickson Mound, Fulton County, Illinois." *Proceedings of the Indiana Academy of Science*. 77 (1968) pp. 102–108.

Bliss, Willard. "Feeding Per Rectum: As Illustrated in the Case of the Late President Garfield and Others." *The Medical Record*. New York, 1882.

Bliven, Bruce Jr. *Under the Guns: New York 1775–1776*. New York: Harper & Row, 1972.

Bobrick, Benson. *Angel in the Whirlwind*. New York: Simon and Schuster, 2011.

Borneman, Walter R. *The French and Indian War*. New York: Harper Collins. 2007.

Bradford, Alden. *History of Massachusetts*. Richardson and Lord, 1825.

Brandt, Allan M. "The Syphilis Epidemic and its Relation to AIDS," *Science* 239 (January 1988), 375–380.

Brown, Fred N. *Rediscovering Vinland: Evidence of Ancient Viking Presence in America*. iUniverse Inc., 2007.

Brumwell, Stephen. *Redcoats: The British Soldier and War in the Americas, 1755–1763*. Cambridge: Cambridge University Press, 2006.

Buikstra, Jane E. *Prehistoric Tuberculosis in America*. Northwestern University Archaeological Program, 1981.

Burrage, Henry S. (editor). *Early English and French Voyages, Chiefly from Hakluyt, 1534–1608*. New York: Charles Scribner's Sons, 1906.

Burrows, Edwin G. "The Prisoners of New York." *Long Island Historical Journal.* Vol. 22, Issue 2, 2011.

Byerly, Carol R. *Good Tuberculosis Men: The Army Medical Department's Struggle with Tuberculosis.* Dept. of the Army, 2014.

Calloway, Colin G. *The Scratch of a Pen—1763 and the Transformation of North America.* New York: Oxford Press, 2006.

Campbell, Ballard C. *Disasters, Accidents, and Crises in American History.* Infobase Publishing, 2008.

Carey, Mathew. *A short account of the malignant fever, lately prevalent in Philadelphia :with a statement of the proceedings that took place on the subject, in different parts of the United States, to which are added, accounts of the plague in London and Marseilles, and a list of the dead from August 1, to the middle of December, 1793.*

Carus, W. Seth. "Bioterrorism and Biocrimes: The Illicit Use of Biological Agents Since 1900." Washington DC: Center for Counterproliferation Research, National Defense University, 1998.

Carroll, Jennifer Lee. *The Speckled Monster: A Historical Tale of Battling Smallpox.* New York:Plume, 2004.

Cartier, Jacques. *A Shorte and Briefe Narration.*

Carus, W. Seth. "Bioterrorism and Biocrimes: The Illicit Use of Biological Agents Since 1900." Washignton DC: Center for Counterproliferation Research, National Defense University, 1998.

Cary, Archibald. *Archibald Cary to R. H. Lee, December 24, 1775.* Letter.

Cave, Alfred. *The French and Indian War.* Greenwood Publishing Group, 2004.

Chadwick, James Read. "Dr. Johann David Schoepf, Surgeon of the Anspach-Bayreuth Troops in America, 1777–1784" *Medical Library and Historical Journal.* 1905 July; 3(3): 157–165.

Chamberlain, Neal. "Biological Warfare/Bio/Terrorism Handbook." http://www.atsu.edu/faculty/chamberlain/bioterror/index.htm. Accessed October 10, 2012.

Chernow, Ron. *Alexander Hamilton.* New York City: Penguin Press, 2004.

Chernow, Ron. *Washington: A Life.* New York City: Penguin Press, 2010.

Cirillo, Vincent J., *Bullets and Bacilli: The Spanish-American War and Military Medicine.* H-War, H-Net Reviews. April, 2004.

Claiborne, William. *William C.C. Claiborne to James Scurlock, June 2, 1802.* Letter.

Clinton, Henry. *The American Rebellion.*, ed. William Willcox New Haven: Yale University Press, 1954.

Coates, John. *Internal Medicine in WWII.* Washington DC: Office of the Surgeon, 1963.

Coffin, Charles. *The Life and Services of Major Gen. John Thomas.* NY: Egbert, Hovey, and King, 1844.

Cohen, Mark Nathan. *Health and the Rise of Civilization.* New Haven: Yale University Press, 1991.

Connor, Seymour V. *North America Divided: The Mexican War 1846–1848.* NY: Oxford Press, 1971.

Cook, Noble D. *Born to Die: Disease and New World Conquest, 1492–1650.* Cambridge: Cambridge University Press, 1998.

Cook, Noble D. *Demographic Collapse: Indian Peru, 1520–1620.* Cambridge University Press, 1982.

Coppolino, Carl. "Lee's Illness Lost Gettysburg" Gatehouse Press (Jan. 26, 2012).

Cornwallis, Charles. *Lord Cornwallis to Sir Henry Clinton, April 10, 1781.* Letter.

Cornwallis, Charles. *Lord Cornwallis to Lord George Germain, Sept. 19, 1780.* Letter.

Cortés, Hernan. *Five Letters.* New York: W.W. Norton & Company, 1962.

Cox, T.M. "King George III and porphyria: an elemental hypothesis and investigation." *Lancet* 366 (2005): 332–335.

Croddy, Eric. "Tularemia, Biological Warfare, and the Battle for Stalingrad." *Military Medicine*, Vol. 166, No. 10 (Oct. 2001).

Crosby, Alfred W. *The Columbian Exchange: Biological and Cultural Consequences of 1492.* Westport, CT: Greenwood, 1973.

Crosby, Alfred W. *Germs, Seeds & Animals: Studies in Ecological History.* M.E. Sharpe, 1994.

Crump, Jennifer. *Canada under Attack.* Toronto: Dundam, 2010.

Cunningham, H.H. *Doctors in Grey: The Confederate Medical Service.* Gloucester, Ma: Peter Smith, 1970.

Currie, William. *A Description of the Malignant, Infectious Fever Prevailing at Present in Philadelphia With an Account of the Means to Prevent Infection, and the Remedies and Method of Treatment, Which Have Been Found Most Successful.* Printed by T. Dobson, No. 41, South Second-Street, 1793.

Davidson, Jonathan. *Downing Street Blues: A History of Depression and Other Mental Afflictions in British Prime Ministers.* New York: McFarland, 2010.

Davis, Graeme. *Vikings in America.* Edinburgh: Birlinn Publishers, 2009.

Davis, William C. *Battle at Bull Run: A History of the First Major Campaign of the Civil War.* Louisiana State University Press, 1981.

De Bevoise, Ken. *Agents of Apocalypse: Epidemic Disease in the Colonial Philippines.* Princeton University Press, 1995.

De Galvez, Jose. *Jose de Galvez to Luis de Unzaga, Dec. 24, 1776.* Letter.

De Oviedo y Valdes, Gonzalo Fernandez. *Writings from the Edge of the World: The*

Memoirs of Darien, 1514–1527. University of Michigan Press, 2008.

De Oviedo y Valdes, Gonzalo Fernandez. *Natural History of the West Indies.* University of North Carolina Press, 1959.

De Roo, Peter. *History of America Before Columbus.* Nabu Press, 2011.

de Vaca, Cabeza. *Chronicles of the Narvaez Expedition.* NY: W.W. Norton and Company, 2013.

Devèze, Jean. *An Enquiry Into, and Observations Upon the Causes and Effects of the Epidemic Disease, which Raged in Philadelphia from the Month of August Till towards the Middle of December, 1793.* Philadelphia: Printed by Parent, 1794.

DeWitt, Benjamin. *An Account of the Internment of the Remains of 11,500 American Seamen, Soldiers, and Citizens, Who Fell Victims to the Cruelties of the British on board their Prison Ships at the Wallabout During the American Revolution.* (1808).

Diamond, Jared. *Guns, Germs, and Steel.* New York City: W.W. Norton and Company, 2005.

Dick, Elisha C. "Facts and Observations Relative to the Disease of Cynanche Trachealis, or Croup." *Philadelphia Medical and Physical Journal.* Letter. October 7, 1808, Published May, 1809—Page 253 (supplement).

Dobson, M.J. "History of Malaria in England." *Journal of the Royal Society of Medicine.* (1989) 82 (Supp. 17): 3–7.

Dobyns, Henry F. *Their Number Become Thinned—Native American Population.* Knoxville: University of Tennessee Press, 1983.

Donkin, Major Robert. *Military Collections and Remarks.* New York: H. Gaines, 1777.

Douglas, John. *Medical Topography of Upper Canada.* London: Burgis and Hill, 1819.

Dowdey, Clifford. *The Wartime Papers of Robert E. Lee.* Boston: Little, Brown, & Company, 1961.

Drake, James David. *King Philip's War: Civil War in New England.* University of Massachusetts Press, 1999.

Dull, Jonathan R. *The Age of the Ship of the Line: The British and French Navies, 1650–1815.* Lincoln: University of Nebraska Press, 2009.

Duncan, Louis C. "Medical Men in the American Revolution: The New York Campaign of 1776." *New York Medical Journal* Vol. CXII No. 11, Sept. 11, 1920.

Dynamics in Eastern North America. Knoxville: University of Tennessee Press, 1983.

Earle, Alice Morse. *Customs and Fashions in Old New England.* 1893.

Eddy, Mary Baker. "Contagion" in *Miscellaneous Writings* (1896) http://www.gutenberg.org/files/31427/31427-h/31427-h.html

Fanselow, Julie. *Traveling Lewis and Clark Trail*. Guilford, CT: Globe Pequot Press, 2007.

Ferling, John E. *The First of Men: A Life of George Washington*. Knoxville: University of Tennessee Press, 1988.

Fischer, David H. *Washington's Crossing*. New York: Oxford University Press, 2006.

Fitzgerald, Michael. "Did 'Stonewall' Jackson Have Asperger's Syndrome?" *Society of Clinical Psychologists*. Accessed at http://www.scpnet.com/paper2_2.htm

Flegal, Katherine M. et al. "Prevalence and Trends in Obesity Among US Adults, 1999–2008." JAMA (2010), pp. 234–241.

Fonvielle, Chris E. *The Wilmington Campaign*. Mechanicsburg: Stackpole Books, 2001.

Forbes, Esther. *Paul Revere*. New York: Houghton Mifflin Harcourt, 1999.

Fowler, William H. *Empires at War: The French and Indian War and the Struggle for North America*. Bloomsbury Publishing, 2009.

Franklin, Ben. *The Retort Courteous* (1786).

Fulghum, Neil. "Hugh Walker and North Carolina's 'Smallpox Currency' of 1779." *The Colonial Newsletter* (Dec. 2005) Accessed at http://www.lib.unc.edu/dc/money/1779article.pdf

Gallagher, Gary W. ed. *Antietam: Essays on the 1862 Maryland Campaign*. Kent State University Press, 1989.

Gelston, Arthur. "Typhus Fever: Report of an Epidemic in New York City in 1847." *The Journal of Infectious Disease* Vol. 136, No. 6 (Dec. 1977), pp 813–821.

Gómez i Prat, Jordi et al. "Prehistoric Tuberculosis in America: Adding Comments to a Literature Review." Mem Inst Oswaldo Cruz, Rio de Janeiro, Vol. 98(Suppl. I) (2003): 151–159. http://memorias.ioc.fiocruz.br/98sup/6p.pdf

Grant, Ulysses S. *Personal Memoires of US Grant*. Gutenberg eBook, 2004.

Grayson, Cary T. *Cary T. Grayson to Samuel Ross, April 14, 1919*. Letter.

Haefeli, Evan and Sweeney, Kevin. *Captive Histories: English, French, and Native Narratives of the 1704 Deerfield Raid*. University of Massachusetts Press, 2006.

Haefeli, Evan and Sweeney, Kevin. *Captors and Captives: The 1704 French and Indian Raid on Deerfield*. University of Massachusetts Press, 2005.

Harrell, Laura D. S. "Preventive Medicine in the Mississippi Territory, 1799–1802." *Bulletin of the History of Medicine* (1966) 40(4): 364–37.

Harriot, Thomas. *A Brief and True Report of the New Found Land of Virginia*. 1588

Harrison, Robert. *Robert H. Harrison to Council of Massachusetts, December 3, 1775.* Letter.

Hearn, Chester G. *Lincoln and McClellan at War.* LSU Press, 2012.

Hedley-Whyte, John and Debra R. Milamed. "Surgical Travellers: Tapestry to Bayeux," *Ulster Medical Journal,* 83 (3), 171–177.

Henry, Alexander. *Alexander Henry's Travels and Adventures in the Years 1760–1776.* Chicago: R.R. Donnelly and Sons, 1921.

Hickey, Donald. *The War of 1812.* Chicago: University of Illinois Press, 1989.

Hopkins, Donald R. *The Greatest Killer: Smallpox in History.* Chicago: University of Chicago Press, 2002.

Horn, James. *A Kingdom Strange: The Brief and Tragic History of the Lost Colony of Roanoke.* Basic Books, 2011.

Humbert, Jules. *Loccupation allemande do Venezuela.* 1905.

Humphrey, David. *Life of General Washington.* Atlanta: University of Georgia Press, 2006.

Hunter, Douglas. *The Race to the New World: Christopher Columbus, John Cabot, and the Lost History of Discovery.* New York: Palgrave Macmillan, 2011.

Hurt, R. Douglas. *Agriculture and the Confederacy: Policy, Productivity, and Power in the Civil War South.* UNC Press Books, 2015.

Hutson, Richard. *Richard Hutson to Isaac Hayne, May 27, 1776.* Letter. Richard Hutson Letterbook, Langdon Cheves III Papers: Extracts from Private Journals.

Ingram, Daniel Patrick. *In the Pale's Shadow: Indians and British Forts in Eighteenth Century America.* UMI Dissertation Publishing, 2011.

Jefferson, Thomas. *Thomas Jefferson to Dupont de Nemours, July 1807.* Letter. *The Jefferson Cyclopedia.* Funk and Wagnells, 1907.

Jefferson, Thomas. *Thomas Jefferson to William Claiborne, July 7, 1804.* Letter.

Jennings, Francs. *The Invasion of America: Indians, Colonialism, and the Cant of Conquest.* New York: Norton, 1975.

Jones, Absalom. *A Narrative of the Proceedings of the Black People, During the Late Awful Calamity in Philadelphia, in the Year 1793: and a Refutation of Some Censures, Thrown Upon Them in Some Late Publications.* Philadelphia: Printed for the authors, by William W. Woodward, at Franklin's Head, no. 41, Chesnut-Street, 1794.

Jortner, Adam. "Cholera, Christ, and Jackson: The Epidemic of 1832 and the Origins of Christian Politics in Antebellum." *Journal of the Early Republic* (Summer 2007).

Kale, David. *The Boston Harbor Islands: A History of an Urban Wilderness.* (Charleston: History Press, 2007).

Keevil, J.J. *Medicine in the Navy, Vol. 1.* Livingston, 1957.

Kendrew, W.G. *The Climates of the Continents.* Oxford: Clarendon Press, 1937.

Kenton, Edna. *Jesuit Relations and Allied Documents.* New York: Kessinger, 2010.

King, Horatio. *Turning on the Light: A Dispassionate Survey of President Buchanan Administration.* University of California Libraries, 1895.

Kiple, Kenneth F. et al.. *Biological Consequences of the European Expansion.* New York: Ashgate, 1997.

Kiple, Kenneth F. *Plague, Pox, & Pestilence.* New York: Barnes & Noble Books, 1997.

Kiracofe, James B. *A Case of Mistaken Identity! Leprosy, Measles, or Smallpox? Old World Names for a New World Disease: Bartonellosis.* A paper presented at the 118th Annual Meeting of the American Historical Association. January 9, 2004. Washington DC.

Knox, Henry. *Henry Knox to George Washington, Oct. 23, 1786.* Letter.

Kohn, George C. *Encyclopedia of Plague and Pestilence: From Ancient Times to the Present.* New York: Checkmark Books, 2002.

Krauss, Hartmut. *Zoonoses: Infectious Diseases Transmissible from Animal to Humans.* ASM Press, 2003.

Krohn, Katherine E. *Wild West Women.* New York: Lerner, 2005.

Krupat, Arnold. *'That the People Might Live': Loss and Renewal in Native American Elegy.* Cornell University Press, 2012.

Kukla, Jon. *A Wilderness So Immense.* NY: Alfred A. Knopf, 2003.

Lamb, Martha. *History of the City of New York: Its Origin, Rise, and Progress,* Volume I. Cosmo Inc., 2005.

Lamers, William Mathias. *The Edge of Glory: A Biography of William S. Rosecrans.* LSU Press, 1999.

Leach, Douglas Edward. *Arms for Empire: Military History of the British Colonies in North America, 1607–1763.* Parsippany NJ: Collier Macmillian Ltd., 1974.

Leclerc, Charles. *Charles Leclerc to Napoleon, Oct. 2, 1802.* Letter.

Lee, Henry. *General Lee to Congress, February 1776.* Letter.

Lefkowitz, Arthur. *Benedict Arnold's Army.* Savas Beatie, 2008.

Lesser, Charles H. *The Sinews of Industry: Monthly Strength Reports of the Continental Army.* Chicago: 1976.

Lewis, Michael. *The Spanish Armada.* London: Pan Books Ltd., 1960.

Lind, James. *James Lind to Thomas Pennant Aug. 1793.* Letter.

Lind, James. *Two Papers on Fevers and Infection Which Were Read Before the Philosophy and Medical Society in Edinburgh.* (1763).

Lind, James. *A Treatise on the Scurvy: In Three Parts, Containing an Inquiry Into the Nature, Causes, an Cure, of that Disease, Together with a Critical and Chronological View of what Has Been Published on the Subject* (1772).

Loge, Ronald V. " 'Two dozes of barks and opium': Lewis and Clark as Physicians." *We Proceeded On,* Vol. 23, No. 1 (February 1997), 10–15, 30. Reprinted from *The Pharos of Alpha Omega Alpha,* Vol. 59, No. 3 (Summer, 1996), 26–31.

Lowry, Thomas Power. *The Stories the Soldiers Wouldn't Tell: Sex in the Civil War.* Stackpole Books, 2012.

Lynnerup, Niels. "The Greenland Norse and the Little Ice Age: Large Scale Climate Changes to a Small Scale Society." 2009 IOP Conference Series: Earth and Environmental Science. Vol. 6 Session 7.

Macalpine, I. "The Insanity of King George III: A Classic Case of Porphyria." *British Medical Journal.* 1966, 1 (5479): 65–71.

MacMahon, Edward B. *Medical Cover-ups in the White House.* Washington DC: Farragut, 1987.

Madison, James. *James Madison to Henry Dearborn, August 8, 1813.* Letter.

Mahon, John K. *The War of 1812.* Gainesville: University of Florida Press, 1972.

Mallon, Mary. *Mary Mallon to George Francis O'Neill, June 1909.* Letter.

Mainwaring, R.D. et al., "The Cardiac Illness of Robert E. Lee," *Surg. Gynecol. Obstet.* Vol. 174, No. 3 (March 1992), 237–244.

Mandell, Daniel R. *King Philip's War: The Conflict Over New England.* Infobase Publishing, 2009.

Mann, James. *Medical sketches of the campaigns of 1812, 13, 14.* H. Mann & Co, 1816.

Manning, Charles and Merrill Moore. "Sassafras and Syphillis." *The New England Quarterly* Vol. 9, No. 3 (Sept. 1936), pp. 473–475.

Mao Zedong, *Mao Zedong to I. V. Stalin, Feb. 21, 1951.* Telegram.

Massry, Shaul G et al. "History of Nephrology." *The American Journal of Nephrology* 1997:17. p 233–240.

Masterson, Karen. *The Malaria Project.* New York: New American Library, 2014.

Mather, Cotton. *Magnalia Christi Americana.*

Mattes, Merrill. *The Great Platte River Road.* Lincoln: University of Nebraska Press, 1987.

McCaa, Robert et al. "Why Blame Smallpox? The Death of the Inca Huayna Capac and the Demographic Destruction of Tawantinsuyu." American Historical Association Annual Meeting (Jan. 8–11, 2004).

McCandless, Peter. *Slavery, Disease, and Suffering in the Southern Lowcountry.*

Cambridge: Cambridge University Press, 2011.

McDonnell, Sharon. "Revolutionary Martyrs." *American Spirit*. March/April 2007.

McKean, Brenda Chambers. *Blood and War at my Doorstep* Vol. 2. Xlibris Corporation, 2011.

Metchnikoff, Elie. *The New Hygiene: Three Lectures on the Prevention of Infectious Disease*. Chicago: W.T. Keener and Company, 1907.

Minutes of the Proceedings of the Committee Appointed on the 14th September, 1793, by the Citizens of Philadelphia, the Northern Liberties and the District of Southwark, to Attend to and Alleviate the Suffering of the Afflicted with the Malignant Fever, Prevalent in the City and Its Vicinity, With an Appendix. Philadelphia: Printed by R. Aitken & Son, 1794.

Mires, Peter B. "Contact and Contagion: The Roanoke Colony and Influenza." *Historical Archaeology*, Vol. 28 No 3 (1994).

Mockford, Jim. "The Mystery of the Brig *Owhyhee*'s Anchor and the Disappearance of Captain John Dominis." *The Northern Mariner* XVIII, Nos. 3–4 (July–Oct. 2008), 105–118.

Moses, Bernard. *The Spanish Dependencies in South America*. Vol. I of Library of Latin American History and Culture. Cooper Square Publishers, 1966.

Nathanson, Constance. *"Disease Prevention as Social Change*. Russell Sage Foundation, 2009.

National Tuberculosis Association. *The Modern Health Crusade: A National Program of Health Instruction*. 1922.

Norton, Mary Beth. *In the Devil's Snare: The Salem Witchcraft Crisis of 1692*. New York: Alfred A. Knopf, 2000.

Oberg, Michael Leroy. *Dominion and Civility: English Imperialism and Native America 1595–1685*. Cornell University Press, 2004.

O'Flaherty, Daniel. *General Jo Shelby: Undefeated Rebel*. UNC Press Books, 2000.

Olch, Peter D. "Treading the Elephant's Tail: Medical Problems on the Overland Trails." *Overland Journal*, Vol 6, Num 1 (1988).

Omi, M. and H. Winant. *Racial formation in the United States : from the 1960s to the 1980s*. New York: Routledge & Kegan Paul, 1986.

Ott, Thomas O. *The Haitian Revolution, 1789–1804*. University of Tennessee Press, 1973.

Oviatt, Edwin. *The Beginning of Yale 1701–1726*. New Haven: Yale University Press, 1916.

Packard, Frances R. *History of Medicine in the United States*, Vol. 1. NY: Hafner Publishers, 1932.

Parascandola, John. *Sex, Sin, and Science: A History of Syphilis in America*. New

York: Greenwood, 2008.

Petriello, David R. "Yellow Fever, Yellow Peril: Disease and Nativism in American History." presented before the SAHMS Conference in Jackson, MS (March 2015).

Pinals, Robert S. "John Hancock's Gout." *Journal of Clinical Rheumatology*. Vol. 18, Issue 4(June 2012) pp 217–219.

Plummer, Brenda Gayle. *Haiti and the United States: Psychological Movement*. University of Georgia Press, 1992.

Pollo, Jose Toribio. "Apuntes sobre las epidemias del Peru." *Revista Historica*, 5(1913):50–109.

"Popp's Journal." *Pennsylvania Magazine of History and Biography* Vol. 26 (1902)

Prevost, Augustine. *Augustine Prevost to Sir Henry Clinton, July 14 and July 30, 1779*. Letters.

Proceedings of the College of Physicians of Philadelphia, Relative to the Prevention of the Introduction and Spreading of Contagious Diseases. Philadelphia: Printed by Thomas Dobson, 1798.

Pybus, Cassandra. "Jefferson's Faulty Math: The Question of Slave Defections in the American Revolution." *The William and Mary Quarterly*. Third Series, Vol. 62, No. 2 . (April 2005), 243–264.

Quimby, Roger Sherman. *The US Army in the War of 1812: An Operational and Command Study*. Michigan: Michigan University Press, 1997.

A Report of the Record Commissioners of the City of Boston: Containing the Selectmen's Minutes from 1764 to 1768. Boston: Heritage Books, 1889.

Reprint from the Public Health Reports, Issues 285–326, United States Public Health Service.

Rhoades, E. *American Indian Health*. Baltimore: Johns Hopkins University Press, 2000.

Richter, Daniel K. "War and Culture: The Iroquois Experience" *The William and Mary Quarterly*, 3rd Ser., Vol. 40, No. 4. (Oct., 1983), pp. 528–559.

Robertson, R.C. *Rotting Face: Smallpox and the American Indian*. Cauton Press, Caldwell Idaho 2001.

Robins, E.A. *Salem Witchcraft and Hawthorne's House of the Seven Gables*. Westminster MD: Heritage Books, 2007.

Roosevelt, Theodore. *The Rough Riders*. Desert Publications, 1996.

Rosenberg, Charles E. *The Cholera Years: The United States in 1832, 1849,and 1866*. Chicago: The University of Chicago Press, 1962.

Ross, Charles. *The Correspondence of Charles, First Marquis Cornwallis*. 1. 3 vols. London: 1859.

Rotberg, Robert I. *Health and Disease in Human History: A Journal of Interdisciplinary History Reader.* MIT Press, 2000.

Ruby, Robert H. *The Chinook Indians: Traders of the Lower Columbia River.* University of Oklahoma Press, 1987.

Rush, Benjamin. *Benjamin Rush to George Washington, Dec. 26, 1777.* Letter.

Rush, Benjamin. *Dr. Benjamin Rush to John Adams, July 4, 1798.* Letter.

Rush, Benjamin. *Observations Upon the Origin of the Malignant Bilious, or Yellow Fever in Philadelphia, and Upon the Means of Preventing It: Addressed to the Citizens of Philadelphia.* Philadelphia: Printed by Budd and Bartram, for Thomas Dobson, at the Stone House, No. 41, South Second Street, 1799.

Rush, Benjamin. *The Autobiography of Benjamin Rush.* New York: Praeger, 1970.

Sandler, Stanley ed. *World War II: An Encyclopedia.* New York: Routledge, 2003.

Satin, Morton. *Death in the Pot: The Impact of Food Poisoning on History.* Amherst: Prometheus Books, 2007.

Sawyer, Frederic H. *The Inhabitants of the Philippines.* New York: C. Scribner and Sons, 1900.

Schlessinger, A.M. *A History of American Life.* New York:Scribner, 1996.

Schoepf, J.D. *Climate and Diseases of America.* Boston: H.O. Houghton & Co., 1875.

Senter, Isaac. *The journal of Isaac Senter, physician and surgeon to the troops detached from the American Army encamped at Cambridge, Mass., on a secret expedition against Quebec, under the command of Col. Benedict Arnold, in September, 1775.* http://archive.org/details/journalisaacsenter00sentrich

Shecut, John L. E. W. *Medical and Philosophical Essays.* Charleston:1819.

Shrewsbury, J.F.D. *A History of Bubonic Plague in the British Isles.* Cambridge: Cambridge University Press, 1970.

Shultz, Suzanne. "Epidemics in Colonial Philadelphia." *Early American Review* (Winter/Spring 2007).

Shyrock, Richard Harrison. *Medicine and Society in America 1660–1860.* Ithaca: Cornell University Press, 1972.

Simpson, Howard N. *Invisible Armies: The Impact of Disease on American History.* Indianapolis: Bobbs-Merrill Company, 1980.

Singer, Barnett et al. *Cultured Force: Makers and Defenders of the French Colonial Empire.* Madison: University of Wisconsin Press, 2008.

Skeen, Carl Edward. *Citizen Soldiers in the War of 1812.* University Press of Kentucky, 1999.

Smillie, Wilson George. *Public Health, Its Promise for the Future.* New York: Macmillan, 1955.

Smith, Beverly C. "The Last Illness and Death of Thomas Jonathan (Stonewall) Jackson," VMI Annual Review (1975), accessed at http://www.vmi.edu/uploadedFiles/Archives/Jackson/Death_and_Funeral/The%20Last%20Illness%20and%20Death%20of%20General%20Jackson.pdf

Smith, Daniel Blake. "Mortality and Family in the Colonial Chesapeake." *Journal of Interdisciplinary History*. 8.3 (Winter, 1978): 403–427

Smith, John. *The General History of Virginia*. NY: Applewood Books, 2006.

Smith, Josiah. *Josiah Smith to James Poyas, Dec. 5, 1780*, "Josiah Smith Letter Book, 1771–1784," Southern Historical Collection, University of North Carolina at Chapel Hill.

Smith, Zachariah Frederick. *The History of Kentucky: From its Earliest Discovery and Settlement to its Historic Characters*. Kentucky: Courier Company, 1892.

Snowden, Frank. *The Conquest of Malaria: Italy, 1900–1962*. New Haven: Yale, 2008.

Soper, F.L. et al. "Typhus Fever in Italy, 1943–1945, and its Control with Louse Powder." *The American Journal of Hygiene* Vol. 45, No. 3 (May 1947), 305–334.

Spelman, Henry. *Relation of Virginia*. Chiswick Press, 1872.

Stearns, Samuel. *An Account of the Terrible Effects of the Pestilential Infection in the City of Philadelphia; With an Elegy on the Deaths of the People; also, A Song of Praise and Thanksgiving, Composed for Those Who Have Recovered, After Having Been Smitten with that Dreadful Contagion*. Providence: Printed for William Child, 1793.

Sternberg, Martha L. *George Miller Sternberg: A Biography*. American Medical Association, 1920.

Stokes, John H. *The Third Great Plague: A Discussion of Syphilis for Everyday People*. Philadelphia: W.B. Saunders Company, 1920.

Stone, Edward T. "The Man Behind Columbus." *American Heritage* Vol. 27, Issue 6 (1976) http://www.americanheritage.com/content/man-behind-columbus

Strode, George K. *Yellow Fever*. NY: McGraw Hill, 1951.

Sullivan, John. *John Sullivan to Philip Schuyler, June 19, 1776*. Letter.

Swiderski, Richard M. *Anthrax: A History*. McFarland and Co. Inc. Publishing, 2004.

Taft, William H. *William H. Taft to A. L. Lowell, Oct. 5, 1919*. Letter.

Taylor, Alan. *The Civil War of 1812*. New York: Alfred A. Knopf, 2010.

Thayer, Theodore. *Nathanael Greene: Strategist of the American Revolution*. New York: Twayne, 1960.

Tomblin, Barbara Brooks. *G. I. Nightingales: The Army Nurse Corps in World War*

II. University Press of Kentucky, 2003.

Trent, William. *"Diary of William Trent"* (June 24, 1763).

Trigger, Bruce. *Natives and Newcomers: Canada's Heroic Age Reconsidered.* Toronto: McGill Queens University Press, 1986.

Tucker, Jonathan B. *Scourge: The Once and Future Threat of Smallpox.* New York: Grove Press, 2002.

A True Declaration of the Estate of the Colonie in Virginia. Washington: Peter Force, 1844.

U.S. Marine Hospital Service. *Annual Report of the Supervising Surgeon General of the Marine Hospital.* US Government Printing Office, 1896.

Wafer, Lionel. *A New Voyage and Description of the Isthmus of America.* New York: The Burrows Brother Company, 1903.

Warren, Martin. *Purple Secret: Genes, "Madness," and the Royal Houses of Europe.* Corgi Books, 1999.

Washington, George. *George Washington to James McHenry, July 3, 1789.* Letter.

Washington, George. *George Washington to John Hancock, Sept. 8, 1776.* Letter.

Washington, George. *George Washington to Tobias Lear, Sept. 10, 1793.* Letter.

Washington, George. *George Washington to Henry Knox, Sept. 9, 1793.* Letter.

Washington, George. *George Washington to John Adams, Sept. 22, 1776.* Letter.

Washington, George. *George Washington to William Shippen, January 6, 1777.* Letter

Washington, George. *George Washington to Sir William Howe, January 13, 1777.* Letter.

Washington, George. *George Washington to Alexander Hamilton, Sept. 6, 1793.* Papers of Alexander Hamilton. Vol. 15 p325, in ed. Note.

Washington, George. *George Washington to President of Congress, July 21, 1775.* Letter.

Washington, George. *George Washington to John Augustine Washington, July 27, 1775.* Letter.

Washington, George. *George Washington to General Gage, August 11, 1775.* Letter.

"The Papers of George Washington." University of Virginia. Number Five (Spring 2002).

Watkins, Walter Kendall. *Soldiers in the Expedition to Canada in 1690 and Grantees of the Canada Townships* (1898).

Westmorland, J.G. et al. *Atlanta Medical and Surgical Journal* (1867), 242.

Whiteley, Peter. *Lord North: The Prime Minister Who Lost America.* New York: Continuum Press, 2007.

Willison, G.F. *Saints and Strangers.* New York: Reynal and Hitchcock, 1945.

Winthrop, Fitz-John. *Fitz-John Winthrop to the Governor and Council of Connecti-*

cut. July 29, 1690. Letter.

Winthrop, Fitz-John. "Journal of the Expedition to Canada."

Winthrop, Fitz-John. *Fitz-John Winthrop to the Governor and Council of Connecticut. Aug. 15, 1690*. Letter.

Wise, John. *Two Narratives of the Expedition Against Quebec, A.D. 1690, Under Sir William Phips*. University of Michigan Library, 1902.

Withington, Lothrop, ed. *The Diary of Caleb Haskell*. Newburyport: William H. Huse and Company, 1881.

Wood, Betty. *The Origins of American Slavery*. Hill and Wang., NY, 1997.

Wood, P. H. "Sickness and settlement: disease as factor in early colonization of New England." *The Pharos of Alpha Omega Alpha* 27, 1964: 98–101.

Wright, Conrad Edick. "John Harvard: Brief life of a Puritan philanthropist." *Harvard Magazine*. (Jan.–Feb. 2000)

INDEX

Adams, John, 91, 95, 96, 102, 130, 133
Alaska, 16, 215
Aleutian Island Campaign, 207
Alger, Russell, 175, 177, 178, 179
Algonquin, 30, 43, 44, 52
Allen, Ethan, 98
Al-Qaeda, 220–224
Andersonville, 156
angina pectoris, 133
Anson, George, 56
anthrax, 188, 189, 191, 210–213,
    215–217, 219–222
Arnold, Benedict, 88–89
Aztec, 11–14, 20, 31

Bacon's Rebellion, 33
Baldwin Report, 212
bartonellosis, 15, 17
Battle of Buena Vista, 144
Battle of Bull Run, First, 157
Battle of Camden, 111–112
Battle of Corinth, 160
Battle of Crysler's Farm, 134
Battle of El Alamein, 202
Battle of Fallen Timbers, 122
Battle of Gettysburg, 164–166
Battle of Guadalcanal, 207
Battle of King's Mountain, 112
Battle of Long Island, 93
Battle of Mexico City, 144
Battle of San Juan Hill, 177
Battle of Santiago, 177
Battle of Saratoga, 108
Battle of Savannah, 109

Battle of Stalingrad, 205
Battle of Stono Ferry, 109
Battle of the North Anna River, 167
Battle of the Plains of Abraham, 69
Battle of Tippecanoe, 129
Battle of Veracruz, 144–146
Battle of Vicksburg, 162
Battle of Yorktown, 112–113
Beaver Wars, 42
bin Laden, Osama, 220–221
biological weapons, 19, 113, 189, 205,
    210–213, 215–217, 219, 221–223
Black Assize, 28
Black Hawk War of 1832, 142
Black Legend, 12, 20
Blackburn, Luke P., 168–169
Blue, Rupert, 191
Boston, 49–52, 75–77, 80–85, 88, 90,
    93, 96, 102, 104, 113, 138, 173, 195
Boston, Siege of, 80
Braddock, Edward, 65–67
bubonic plague, 182, 200, 211, 216,
    222–223
Buchanan, James, 152–153
Buckongahelas, 129
Buffalo Fever, 135

California, 11, 139, 143, 146–148, 188,
    194
Calomel Rebellion, 170
Cayuse War, 147–148
Chamberlain-Kahn Act, 192
Charleston, 53, 54, 77, 97, 109, 110,
    157, 164

Charles Town. *See* Charleston
cholera, 20, 138, 142, 143, 147–149,
    168, 182, 185, 191, 211, 216, 217
Church, Benjamin, 85
Civil War, 148, 149, 151–166
Clinton, Henry, 97, 108–110, 112
Cold War, 211–218
Connecticut War, 39
Cornwallis, Charles, 110–113
Cortes, Hernan, 11–13, 19–20, 70
Cuba, 13, 53, 61, 68, 70, 174–179, 182,
    193, 216, 220

d'Iberville, Pierre Le Moyne, 48, 49, 53
DDT, 203, 204, 206, 208, 209
de Soto, Hernan, 15
Deerfield Massacre, 46
dengue fever, 110, 200, 205, 206,
diarrhea, 44, 82,88, 95, 131, 133, 135,
    154, 163, 166, 167, 197, 208
Dilger, Anton Casimir, 189–190
diphtheria, 36, 192
Dodge Commission, 178
Dodge-Leavenworth Expedition,
    142–143
Dugway Proving Grounds, 210, 213,
    215
Dunmore, Lord, 96, 97, 102
dysentery, 18, 34, 40, 57, 65, 66, 68, 82,
    88, 93, 94, 99, 100, 101, 110, 111,
    113, 126, 128, 131, 133–135, 144,
    151, 153, 154–156, 160, 162, 163,
    179, 186

Ecuyer, Simeon, 71
Equine Influenza Epidemic of 1872,
    173

Fort Detrick, 210, 212–214, 219, 220
Fort Duquesne, 64–66, 71
Fort Edward, 67
Fort Frontenac, 47, 51, 52
Fort George, 133, 134
Fort Niagara, 65

Fort Oswego, 66, 70
Fort Pitt, 71, 72, 75
Fort Ticonderoga, 70, 80, 84, 88
Fort William Henry, 67, 72
Franklin, Benjamin, 67, 112–113
French and Indian War, 46, 62, 64, 69,
    70, 73, 75, 76

Geneva Protocol of 1925, 199
George III, 78
Ghost Dance, 150
glanders, 167, 188, 189, 191, 210
gonorrhea, 127, 128, 167
gout, 77, 114, 116, 119, 120, 121, 176
Grant, Ulysses S., 147, 149, 160, 162,
    163, 166
Guadeloupe, 69, 70, 124

Halleck, Henry, 160, 161
Hamilton, Alexander, 120
Hamilton, John B., 176
Hammond, William A., 169–171
Hancock, John, 91, 95, 118–120
Harmar, Josiah, 120–121
Havana, 53, 68, 69, 92, 181,
Havana Yellow Fever Epidemic, 53, 68
hepatitis, 17, 161, 200, 217
herpes, 17
Hill, A.P., 167
HMS Jersey, 98–99
Huey Cocoliztli, 17
Hull, William, 131
Huron Indian Epidemics, 44
hypomagnesemia, 166

Inca, 14, 15, 17
influenza, 19, 31, 36, 40, 44, 45, 114,
    129, 173, 193–198
Invasion of Canada, 88
Iroquois, 43–45, 47, 49, 52
Ishii, Shiro, 212
ISIS, 223

Jackson, Andrew, 129, 140, 142, 151,

Jackson, Thomas "Stonewall", 164, 165
Jamestown, 29–38
Jefferson, Thomas, 90, 113, 123, 125, 126, 130
Johnson, Andrew, 154–155

King George's War, 61
King Philip's War, 40
King William's War, 47, 48
Korean War, 216, 218

L'Ouverture, Toussaint, 123
Laudanum, 125
Lawson, Thomas, 143–146, 169, 170
Leclerc, Charles, 122–124
Lee, Robert E., 159, 162, 166, 167
Lewis and Clark Expedition, 125–128
Lincoln, Abraham, 155, 159, 162, 168, 169
Lind, James, 58, 59, 84
Little Turtle, 120
Louisbourg, 62, 66, 68
Louisiana Purchase, 122
Lovell, Joseph, 143

Madison, James, 116, 129, 131, 134, 136
Maitland, Thomas, 123
malaria, 13, 33, 57, 58, 60, 72, 82, 88, 91, 92, 97, 109, 110, 113, 120, 122, 126, 134, 136, 139, 154, 156, 161–163, 170, 178, 180, 181, 200–209, 218
Manila, 58, 182, 183
Mathers, Cotton, 52
McClellan, George, 158, 159, 161, 162
mercury, 54, 125, 126, 127, 155, 165, 170
Mexican War, 143–146
Miles, Nelson, 176
Mississippi Culture, 10
Morgan, John, 84, 85
mosquito, 72, 73, 114, 126, 139, 168, 179–181, 187, 204, 205, 209, 214, 215

MYL, 203

Napoleon, 10, 123, 124, 130, 159, 161, 205
National Hotel Illness, 152
New Guinea, 206–208
New York City, 86, 93, 95, 96, 98, 101, 108, 110, 170
Newburgh Conspiracy, 116
North, Lord Frederick, 108, 109, 114
Northwest Indian War, 120, 129

Operation Big Buzz, 214
Operation Big Itch, 214
Operation Big Tom, 215
Operation Dark Winter, 221
Operation Drop Kick, 214
Operation Large Area Coverage, 214
Operation Mongoose, 216
Operation Red Hat, 215
Operation United Assistance, 223
Operation Whitecoat, 214
opium, 125, 128, 132, 135, 155, 165, 170
Oregon, 138, 139, 143, 146, 181, 220

Panama Canal, 179
Peninsula Campaign, 161–162
Pequot War, 38
Perry, Oliver Hazard, 134
Persian Gulf War, 217
Philippine Insurrection, 182
Phips, Sir William, 49–52
Pilgrims, 35–38, 41, 146
Pitt, William, 77
pleuropneumonia, 165
Plymouth, 35–40, 92
pneumonia, 36, 89, 132–134, 140, 165, 190, 193, 194, 201, 204
Pocahontas, 32, 33
Pontiac's Rebellion, 70
porphyria, 78, 79
Porto Bello, 42, 55, 56, 59–61
Potawatomi Trail of Death, 141

Prescott, Samuel, 80
Project Bacchus, 221
Project Jefferson, 221

Q fever, 205, 212, 214,
Quebec, 7, 44, 45, 47, 49–52, 62, 67, 69, 70, 88–91
Queen Anne's War, 52, 53
quinine, 92, 122, 125, 166, 170, 181, 206, 208

Rajneesh, Bhagwan Shree, 220
Reed, Joseph, 83
Reed, Walter, 179, 181, 182
Revere, Paul, 75, 76, 80
Revolutionary Prisoners of War, 98–99
ricin, 192, 193, 212, 222
Roanoke, 29–32
Rockingham, Charles W. Wentworth, 76, 77, 114
Roosevelt, Franklin D., 210–212
Roosevelt, Theodore, 177, 181, 182
Rush, Benjamin, 86, 87, 130
Russia, 10, 16, 105, 191, 200, 203, 205

Sacagawea, 126–127
sandfly fever, 200
Schenectady, 49
Schuyler, John, 50
Schuyler, Philip, 88
Scott, Winfield, 133, 142–146, 159
scurvy, 36, 47–49, 52, 57–59, 62, 66, 68–70, 84, 92, 94, 99, 110, 138, 156, 163
Second Cholera Pandemic, 142
Seven Days Battles, 162
Shafter, William Rufus, 176–177
Shay's Rebellion, 118–119
Sherman, William T., 161
Shippen, William, 86, 87, 103
smallpox, 12–15, 16, 18, 19, 21, 36, 39, 40, 44–46, 50–53, 66, 67, 71, 72, 75–83, 85, 89–93, 96, 97, 99, 101, 103, 104, 113, 115, 129, 131, 141, 143,
146, 148, 155–157, 164, 170, 192, 200, 210, 212, 216
Spanish American War, 154
Spanish Armada, 24–26, 28
Spanish Influenza, 19
Squanto, 36–37
St. Clair, Arthur, 121
St. Jo Program, 213
Stamp Act, 76, 77, 114
Starving Time, 32
Sternberg, George, 178, 179
Surgeon General, See Individual Entries
syphilis, 35, 108, 127, 128, 155, 164, 192, 209

T-2 mycotoxin, 217
Teach, Edward, 54
Tecumseh, 129
Tenskwatawa, 129
tetanus, 160, 190, 192, 200, 217
The Mexican War, 143
Tilton, James, 87, 93, 94, 96, 102, 104, 107, 135
Townshend, Charles, 77
Trail of Tears, 140, 141
Trenton, Battle of, 102, 103
tuberculosis, 17, 36, 40, 80, 118, 148, 149, 204
tularemia, 205, 210, 212, 214, 215
typhoid, 99, 101, 110, 134, 141, 149, 154, 159, 160, 178, 179, 190, 192, 193, 199, 200, 205, 211, 217
typhus, 10, 17, 27, 28, 57–60, 62, 68, 74, 77, 82, 92, 93, 109, 113, 133, 191, 200, 203
Typhus Commission, 203

vaccination, 79, 131, 145, 192, 200, 204, 217
Venereal Disease, 54, 55, 93, 127, 155, 192, 201, 209, 218
Vernon, Edward, 56, 60, 61, 99
Versailles Conference, 196, 198

Vietnam War, 218
von Steinmetz, Erich, 188
von Steuben, Friedrich, 104, 105, 107

War of 1812, 128–137
War of Jenkins Ear, 55
Ward, Artemas, 80
Warren, Joseph, 79, 80, 84,
Washington, George, 65, 81, 91, 102,
    108, 116, 117, 119
Wayne, Anthony, 122
Wilmington Yellow Fever Epidemic,
    163, 164

Wilson, Woodrow, 192, 196–198
Wingina, 30
Winthrop, Fitz-John, 49–51
Wodziwob, 150
Wood, Robert C., 169
World War I, 181, 185–193
Wounded Knee Massacre, 149, 150
World War II, 200–211

Yalta Conference, 212
yellow fever, 13, 20, 42, 52–55, 60, 68,
    99, 123, 124, 136, 144, 145, 163, 164,
    168, 169, 174–181, 212,